普通高等院校"十三五"规划教材——化学化工类

化学化工英语

English for Chemistry and Chemical Engineering

主 编 刘子富

西南交通大学出版社
·成 都·

图书在版编目（CIP）数据

化学化工英语 / 刘子富主编. —成都：西南交通大学出版社，2019.8
普通高等院校"十三五"规划教材. 化学化工类
ISBN 978-7-5643-7036-7

Ⅰ.①化… Ⅱ.①刘… Ⅲ.①化学－英语－高等学校－教材②化学工业－英语－高等学校－教材 Ⅳ.①TQ

中国版本图书馆 CIP 数据核字（2019）第 169674 号

普通高等院校"十三五"规划教材——化学化工类

Huaxue Huagong Yingyu
化 学 化 工 英 语

主编　刘子富

责 任 编 辑	孟　媛
封 面 设 计	何东琳设计工作室
出 版 发 行	西南交通大学出版社 （四川省成都市金牛区二环路北一段 111 号 西南交通大学创新大厦 21 楼）
发行部电话	028-87600564　028-87600533
邮 政 编 码	610031
网　　　址	http://www.xnjdcbs.com
印　　　刷	成都蜀雅印务有限公司
成 品 尺 寸	185 mm × 260 mm
印　　　张	17.75
字　　　数	576 千
版　　　次	2019 年 8 月第 1 版
印　　　次	2019 年 8 月第 1 次
书　　　号	ISBN 978-7-5643-7036-7
定　　　价	48.00 元

课件咨询电话：028-87600533
图书如有印装质量问题　本社负责退换
版权所有　盗版必究　举报电话：028-87600562

Preface

《化学化工英语》是根据教育部 2017 年颁布的《大学英语教学指南》精神要求，旨在进一步贯彻落实高等教育教学改革工作，"满足国家战略需求，为国家改革开放和经济社会发展服务，满足学生专业学习、国际交流、继续深造、工作就业等方面的需要"而为高校大学生提供英语学习方面的积极探索。

根据《大学英语教学指南》的要求，大学英语教学的主要内容可分为通用英语、专门用途英语和跨文化交际三个部分，由此形成相应的三大类课程。然而由于学校类型、层次、生源、办学定位、人才培养目标等方面存在的差异，以增强学生运用英语进行专业和学术交流、从事工作的能力，以提升学生学术和职业素养为目的的专门用途英语课程在大学英语改革工作实践中，受制于课时安排、大学英语教师教育背景和知识结构等方面的影响，教学实践仍在不断的探索中。与此同时，作为一门核心学科（central science），化学在科学研究领域有着举足轻重的地位，也影响到人们生活的方方面面。然而现实生活中因其作为理工类课程之一，在高校学生中的普及程度与其应该得到的效果之间相差甚远。该书通过对化学的简介（第一单元）、化学工程的介绍（第二单元）、化学与化工领域基本概念（第三单元）、分支学科（第四单元）、医药工程（第五单元）、药物学（第六单元）等方面知识的介绍，构建起化学化工领域的基本框架，并通过对原子结构（第七单元）、化学键（第八单元）、化学反应（第九单元）、化学反应影响因素（第十单元）等方面知识的介绍，促进化学化工专业领域基本知识的更大范围的推广和普及。化学实验是化学化工领域学习和研究工作中的一个重要组成部分，所以本书也通过对常用化学实验仪器（第十一单元）、常用实验方法——结晶法（第十二单元）、酸碱滴定法（第十三单元）、咖啡因的分解、提纯、鉴定（第十四单元）等方面知识的介绍，进一步完善和提升学习者对化学化工方面专业知识的了解和认识。此外，本书还通过介绍化学的演进发展（革命）（第十五单元）、俄罗斯在元素周期表中新元素发现方面的努力（第十六单元）、制药工业中化学元素合成的重要价值（第十七单元）、拓展生物催化应用来解决药品研发中所遇到的困难和挑战（第十八单元）等化学化工专业领域相关的权威专业文献资料的赏析，深层透视化学化工专业的发展历程和未来发展走向。该书的附录部分，不仅提供了大量的可供随时查阅的化学化工领域相关专家以及专业词汇的介绍，也通过对英国女王大学（Queen's University）化学化工专业本科留学项目的介绍，帮助学生开阔视野的同时，也为未来出国留学做出些许的准备。

虽然该书在编写过程中得到了安康学院化工学院副教授武立州博士的大量指导和帮助，但由于作者水平有限，难免会存在诸多不足之处或不当之处，恳请各位专家或读者不吝指正。

该书的出版也得到了安康学院教材建设专项基金的资助，作者在此深表感激！

编 者

2019 年 6 月

Contents

Part 1 Chemistry and Chemical Engineering

Unit 1 Chemistry 2
Unit 2 Chemical Engineering 9
   2.1 Etymology 9
   2.2 History of chemical engineering 10
   2.3 New concepts and innovations 11
   2.4 Safety and hazard developments 11
   2.5 Recent progress 11
   2.6 Concepts 12

Part 2 Basic Chemistry

Unit 3 Basic Concepts 16
   3.1 Matter 16
   3.2 Atom 16
   3.3 Element 16
   3.4 Compound 17
   3.5 Substance 17
   3.6 Molecule 18
   3.7 Mole 18
   3.8 Ions and salts 19
   3.9 Acidity and basicity 19
   3.10 Phase 20
   3.11 Redox 21
   3.12 Bonding 21
   3.13 Reaction 22
   3.14 Equilibrium 23
   3.15 Energy 23

Unit 4	Subdisciplines	28
4.1	Analytical chemistry	28
4.2	Biochemistry	29
4.3	Inorganic chemistry	32
4.4	Materials chemistry	33
4.5	Neurochemistry	34
4.6	Nuclear chemistry	35
4.7	Organic chemistry	35
4.8	Physical chemistry	38
4.9	Theoretical chemistry	39
Unit 5	Pharmaceutical Engineering	45
5.1	History	47
5.2	Pharmaceutical industry in the United Kingdom	49
Unit 6	Pharmacy	54
6.1	Disciplines	55
6.2	Pharmacy technicians	55
6.3	Education requirements	56
6.4	History	56
6.5	Practice areas	59
6.6	The future of pharmacy	65
6.7	Symbols	65
Unit 7	Atomic Structure	72
Unit 8	Chemical Bonding	76
8.1	Overview of main types of chemical bonds	77
8.2	History	78
Unit 9	Chemical Reaction	82
9.1	Synthesis	83
9.2	Decomposition	83
9.3	Single replacement	83
9.4	Double replacement	84
9.5	Historical overview	84
9.6	Basic concepts of chemical reactions	85
Unit 10	Factors Influencing the Rate of a Chemical Reaction	90
10.1	Concentration of reactants	90
10.2	Temperature	90
10.3	Medium or state of matter	90

10.4	Presence of catalysts and competitors	90
10.5	Pressure	91
10.6	Mixing	91
10.7	Summary of factors that affect chemical reaction rate	91

Part 3 Laboratory

Unit 11	Laboratory Apparatus: Chemical Instruments	94
Unit 12	Recrystallization	100
12.1	Single-solvent approach	100
12.2	Two-solvent recrystallization	102
Unit 13	Acid-Base Titration	105
Unit 14	Isolation, Purification, and Identification of Caffeine	109
14.1	Background	109
14.2	Procedures	110

Part 4 Academic Reading

Unit 15	A (R)evolution in Chemistry	116
15.1	Enzymes—the sharpest chemical tools of life	117
15.2	Human thought has limitations	117
15.3	Arnold starts to play with evolution	117
15.4	Mating—for more stable evolution	119
15.5	New enzymes produce sustainable biofuel	119
15.6	Smith uses bacteriophages	120
15.7	Bacteriophages—a link between a protein and its unknown gene	120
15.8	Antibodies can fish out the right protein	121
15.9	Antibodies can block disease processes	122
15.10	Winter puts antibodies on the surface of phages	122
15.11	The world's first pharmaceutical based on a human antibody	123
15.12	The start of a new era in chemistry	124
Unit 16	A Storied Russian Lab is Trying to Push the Periodic Table Past Its Limits —and Uncover Exotic New Elements	125
Unit 17	The Importance of Synthetic Chemistry in the Pharmaceutical Industry	134
Unit 18	Extending the Application of Biocatalysis to Meet the Challenges of Drug Development	150

Part 5 Supplementary Reading

Appendix 1: Introduction about the Undergraduate Programme in Queen's University 172

Appendix 2: A List of Chemists 182

Appendix 3: A List of Chemical Engineers 195

Appendix 4: A List of Vocabulary in Chemistry and Chemical Engineering 201

Appendix 5: Element Names and Their Pronunciation 202

Appendix 6: Bilingual List of Common Analytical Instrument and Methods 206

Appendix 7: Terms Used in General and Inorganic Chemistry 209

Appendix 8: Terms Used in Organic and Biological Chemistry 236

Appendix 9: Terms Used in Analytical and Physical Chemistry 245

References 274

Part 1

Chemistry and Chemical Engineering

Unit 1　Chemistry

Chemistry is the science concerned with the composition, structure, and **properties** of matter, as well as the changes it undergoes during chemical reactions.

Chemistry is the study of interactions of chemical substances with one another and energy.

Fig. 1-1　Chemistry experiment.

Chemistry (from Arabic "كيمياء", latinized "chem (kēme)", meaning "value") is the science of matter and the changes it undergoes. The science of matter is also addressed by physics, but while physics takes a more general and fundamental approach, chemistry is more specialized, being concerned with the composition, behavior, structure, and properties of matter, as well as the changes it undergoes during chemical reactions. It is a physical science for studies of various atoms, **molecules**, **crystals** and other **aggregates** of matter whether in isolation or combination, which incorporates the concepts of energy and entropy in relation to the spontaneity of chemical processes.

Disciplines within chemistry are traditionally grouped by the type of matter being studied or the kind of study. These include inorganic chemistry, the study of inorganic matter; organic chemistry, the study of organic matter; biochemistry, the study of substances found in biological organisms; physical chemistry, the energy related studies of chemical systems at macro, molecular and submolecular scales; analytical chemistry, the analysis of material samples to gain an understanding of their chemical composition and structure. Many more specialized disciplines have emerged in recent years, e.g. neurochemistry—the chemical study of the nervous system.

Chemistry is the scientific study of interaction of chemical substances that are constituted of atoms or the subatomic particles: protons, electrons and neutrons. Atoms combine to produce molecules or crystals. Chemistry is often called "the central science" because it connects the other natural sciences such as astronomy, physics, material science, biology, and geology.

The genesis of chemistry can be traced to certain practices, known as alchemy, which had been practiced for several millennia in various parts of the world, particularly the Middle East.

The structure of objects we commonly use and the properties of the matter we commonly interact with, are a consequence of the properties of chemical substances and their interactions. For example, steel is harder than iron because its atoms are bound together in a more rigid crystalline lattice; wood burns or undergoes rapid oxidation because it can react spontaneously with oxygen in a chemical reaction above a certain temperature; sugar and salt dissolve in water because their molecular/ionic properties are such that dissolution is preferred under the ambient conditions.

The transformations that are studied in chemistry are a result of interaction either between different chemical substances or between matter and energy. Traditional chemistry involves study of interactions between substances in a chemistry laboratory using various forms of laboratory glassware.

Ancient Egyptians pioneered the art of synthetic "wet" chemistry up to 4,000 years ago. By 1,000 BC ancient civilizations were using technologies that formed the basis of the various branches of chemistry such as: extracting metal from their ores, making pottery and glazes, fermenting beer and wine, making pigments for cosmetics and painting, extracting chemicals from plants for medicine and perfume, making cheese, dying cloth, tanning leather, rendering fat into soap, making glass, and making alloys like bronze.

The genesis of chemistry can be traced to the widely observed phenomenon of burning that led to metallurgy—the art and science of processing ores to get metals (e.g. metallurgy in ancient India). The greed for gold led to the discovery of the process for its purification, even though the underlying principles were not well understood—it was thought to be a transformation rather than purification. Many scholars in those days thought it reasonable to believe that there exist means for transforming cheaper (base) metals into gold. This gave way to alchemy and the search for the Philosopher's Stone which was believed to bring about such a transformation by mere touch.

Greek atomism dates back to 440 BC, as what might be indicated by the book *De Rerum Natura* (*The Nature of Things*) written by the Roman Lucretius in 50 BC. Much of the early development of purification methods is described by Pliny the Elder in his *Naturalis Historia*.

A tentative outline is as follows:

1. Egyptian alchemy (3,000 BCE – 400 BCE), formulate early "element" theories such as the Ogdoad.

2. Greek alchemy (332 BCE – 642 CE), the Greek king Alexander the Great conquers Egypt and founds Alexandria, having the world's largest library, where scholars and wise men gather to study.

3. Arab alchemy (642 CE – 1200), the Muslim conquest of Egypt (primarily Alexandria); development of the Scientific Method by Alhazen and Jābir ibn Hayyān revolutionise the field of Chemistry.

4. The House of Wisdom (Arabic: بيت الحكمة; Bait al-Hikma), Al-Andalus (Arabic: الأندلس) and Alexandria (Arabic: الإسكندرية) become the world leading institutions where scientists of all religious and ethnic backgrounds worked together in harmony expanding the reaches of Chemistry

in a time known as the Islamic Golden Age.

5. Jābir ibn Hayyān, al-Kindi, al-Razi, al-Biruni and Alhazen continue to dominate the field of Chemistry, mastering it and expanding the boundaries of knowledge and experimentation.

6. European alchemy (1300 – present), Pseudo-Geber builds on Arabic chemistry.

7. Chemistry (1661), Boyle writes his classic chemistry text *The Sceptical Chymist*.

8. Chemistry (1787), Lavoisier writes his classic *Elements of Chemistry*.

9. Chemistry (1803), Dalton publishes his *Atomic Theory*.

The earliest pioneers of Chemistry, and inventors of the modern scientific method, were medieval Arab and Persian scholars. They introduced precise observation and controlled experimentation into the field and discovered numerous Chemical substances.

"Chemistry as a science was almost created by the Muslims; for in this field, where the Greeks (so far as we know) were confined to industrial experience and vague hypothesis, the Saracens introduced precise observation, controlled experiment, and careful records. They invented and named the alembic (al-anbiq), chemically analyzed innumerable substances, composed lapidaries, distinguished alkalis and acids, investigated their affinities, studied and manufactured hundreds of drugs. Alchemy, which the Muslims inherited from Egypt, contributed to chemistry by a thousand incidental discoveries, and by its method, which was the most scientific of all medieval operations."

The most influential Muslim chemists were Geber, al-Kindi, al-Razi, al-Biruni and Alhazen. The works of Geber became more widely known in Europe through Latin translations by a pseudo-Geber in 14th century Spain, who also wrote some of his own books under the pen name "Geber". The contribution of Indian alchemists and metallurgists in the development of chemistry was also quite significant.

The emergence of chemistry in Europe was primarily due to the recurrent incidence of the plague and blights there during the so called Dark Ages. This gave rise to a need for medicines. It was thought that there exists a universal medicine called the Elixir of Life that can cure all diseases, but like the Philosopher's Stone, it was never found.

For some practitioners, alchemy was an intellectual pursuit, over time, they got better at it. Paracelsus (1493–1541), for example, rejected the 4-elemental theory and with only a vague understanding of his chemicals and medicines, formed a hybrid of alchemy and science in what was to be called iatrochemistry. Similarly, the influences of philosophers such as Sir Francis Bacon (1561–1626) and René Descartes (1596–1650), who demanded more rigor in mathematics and in removing bias from scientific observations, led to a scientific revolution. In chemistry, this began with Robert Boyle (1627–1691), who came up with an equation known as Boyle's Law about the characteristics of gaseous state. Chemistry indeed came of age when Antoine Lavoisier (1743–1794), developed the theory of Conservation of mass in 1783; and the development of the Atomic Theory by John Dalton around 1800. The Law of Conservation of Mass resulted in the reformulation of chemistry based on this law and the oxygen theory of combustion, which was largely based on the work of Lavoisier. Lavoisier's fundamental contributions to chemistry were a result of a conscious effort to fit all experiments into the framework of a single theory. He established the consistent use of the chemical balance, used oxygen to overthrow the phlogiston theory, and developed a new system of chemical nomenclature and made contribution to the modern metric system. Lavoisier also worked to translate the archaic and technical language of chemistry into something that could be easily understood by the largely uneducated masses, leading to an increased public interest in chemistry. All these advances in chemistry led to what is usually called the chemical revolution. The contributions of Lavoisier led to what is now called modern chemistry—the chemistry that is studied in educational institutions all over the world. It is because of these and other contributions that Antoine Lavoisier is often celebrated as the "Father of Modern Chemistry". The later discovery of Friedrich Wöhler that many natural substances, organic compounds, can indeed be synthesized in a chemistry laboratory also helped the modern chemistry to mature from its infancy.

The discovery of the chemical elements has a long history from the days of alchemy and culminating in the discovery of the periodic table of the chemical elements by Dmitri Mendeleev (1834–1907) and later discoveries of some synthetic elements.

New words

property ['prɒpətɪ] *n.* 特性，属性；财产，地产；所有权；[戏]道具
molecule ['mɒlɪkjuːl] *n.* 分子；微小颗粒
molecular [mə'lekjələ(r)] *adj.* 分子的，由分子组成的
submolecular [sʌbmə'lekjulə] *adj.* 亚分子的
supermolecular [suːpərmə'lekjulə] *adj.* 超分子的
crystal ['krɪstl] *n.* 结晶（体）；晶体；水晶；水晶饰品
    adj. 水晶的；水晶般的；透明的；清楚的
aggregates ['ægrɪgɪt] *n.* 合计；聚集体；骨料；集料（可成混凝土或修路等用的）
    adj. 总数的，总计的；聚合的；[地]聚成岩的
    vt. 使聚集，使积聚；总计达
incorporate [ɪn'kɔːpəreɪt] *vt.* 组成公司；包含；使混合；使具体化
    vi. 包含；吸收；合并；混合
entropy ['entrəpɪ] *n.* 熵，平均信息量；负熵
spontaneity [ˌspɒntə'neɪətɪ] *n.* 自发性，自然发生；自发行为（行动）
organic [ɔː'gænɪk] *adj.* 有机（体）的；有组织的，系统的；器官的；根本的
inorganic [ˌɪnɔː'gænɪk] *adj.* [化]无机的；无组织结构的；无生物的；无活力的
organism ['ɔːgənɪzəm] *n.* 有机体；生物体；微生物；有机体系，有机组织
biochemistry [ˌbaɪəʊ'kemɪstrɪ] *n.* 生物化学；生物化学成分
analytical [ˌænə'lɪtɪkl] *adj.* 分析的，分析法的；善于分析的
neurochemistry [njʊərə'kemɪstrɪ] *n.* 神经化学
subatomic [ˌsʌbə'tɒmɪk] *adj.* 小于原子的，亚原子的，次原子的

particle ['pɑːtɪkl] *n.* 微粒，颗粒；[数，物]粒子，质点；极小量；小品词
proton ['prəʊtɒn] *n.* [物]质子
electron [ɪ'lektrɒn] *n.* 电子
neutron ['njuːtrɒn] *n.* [物]中子
geoglogy [dʒɪ'ɒlədʒɪ] *n.* 地质学；（某地区的）地质情况；地质学的著作
genesis ['dʒenəsɪs] *n.* <正>创始，起源，发生
practice ['præktɪs] *n.* 练习；实践；（医生或律师的）业务；惯例
　　　　　　 vi. 实行；惯常地进行；练习；实习
　　　　　　 vt. 实行，实践；执业；练习；惯常地进行
alchemy ['ælkəmɪ] *n.* 炼金术；炼丹术；（改变事物、物质的）魔力（方法）；（事物、物质的）神秘变化
millennium [mɪ'lenɪəm] *n.* 一千年；千年期；千禧年
　　　　　　　　复数形式：millennia or millenniums
rigid ['rɪdʒɪd] *adj.* 严格的；僵硬的；（规则、方法等）死板的；刚硬的，顽固的
crystalline ['krɪstəlaɪn] *adj.* 水晶的；似水晶的；结晶质的；清澈的
　　　　　　　　 n. 结晶性，结晶度
lattice ['lætɪs] *n.* 格子框架；类似格子框架的设计
　　　　　　vt. 把……制成格子状；用格子覆盖或装饰
oxidation [ˌɒksɪ'deɪʃn] *n.* 氧化
dissolve [dɪ'zɒlv] *vt.* 使溶解；使液化
　　　　　　 vi. 溶解；融化，液化，分解
dissolution [ˌdɪsə'luːʃn] *n.* 溶解，融化
ionic [aɪ'ɒnɪk] *adj.* 离子的
ambient ['æmbɪənt] *adj.* 周围的，包围着的；环境
preferred [prɪ'fəd] *adj.* 首选的
glassware ['glɑːsweə] *n.* 玻璃器具类
pioneer [ˌpaɪə'nɪə] *n.* 拓荒者；开发者；先驱者；创始者
　　　　　　 vt. 开拓，开发；做（……的）先锋；提倡
synthetic [sɪn'θetɪk] *adj.* 合成的；人造的；模拟的，虚构的
　　　　　　 n. 合成物；合成纤维；合成剂
"wet" chemistry 湿化学（It is a form of analytical chemistry that uses classical methods such as observation to analyze materials. It is called wet chemistry since most analyzing is done in the liquid phase. Wet chemistry is also called bench chemistry since many tests are performed at lab benches.）
pottery ['pɒtərɪ] *n.* 陶器；陶器厂（作坊）；<集合词>陶器类；陶器制造（术）
glaze [gleɪz] *vt.* 装玻璃；上釉于；上光
　　　　　vi.（目光）变得呆滞无神；变得光滑

　　　　　　　　n. 上釉的表面；釉料；光滑面；（浇在糕点上增加光泽的）蛋浆
ferment [fə'ment] *n.* 酶，酵素；发酵剂；骚动，动乱
　　　　　　　　vt. & *vi.* 使发酵；使骚动；酝酿
pigment ['pɪgmənt] *n.* 颜料，色料；[生]色素
　　　　　　　　vt. 给……着色
　　　　　　　　vi. 呈现颜色
cosmetic [kɒz'metɪk] *n.* 化妆品；美发油，发蜡；装饰品；美容术
　　　　　　　　adj. 化妆用的；美容的；装点门面的；表面的
perfume ['pɜːfjuːm] *n.* 香水；香料；香味，香气
　　　　　　　　vt. 使……充满香气；喷香水于……
tan [tæn] *n.* 黄褐色，棕黄色；鞣料；马戏团；晒黑的皮色
　　　　　　　　vt. （使）晒成棕褐色；鞣（革）
　　　　　　　　vi. 晒成棕褐色；
　　　　　　　　adj. 黄褐色的，棕黄色的；鞣皮的
leather ['leðə] *n.* 皮，皮革；皮革制品
　　　　　　　　vt. 用皮革包盖；制成皮，蒙上皮
　　　　　　　　adj. 皮的，皮革的，皮革制的
render ['rendə(r)] *v.* 致使，造成；给予；递交；表达
alloy ['ælɔɪ] *n.* 合金；（合金中的）劣等金属；搀杂品；成色
　　　　　　　　v. 合铸，熔合（金属）；铸成合金；在……中搀以杂质，使（金属）减低成色
bronze [brɒnz] *n.* 青铜；青铜色；铜牌；青铜艺术品
　　　　　　　　adj. 深红棕色的，青铜色的；青铜制的
　　　　　　　　vt. 镀青铜于
　　　　　　　　vi. 变成青铜色，被晒黑
metallurgy [mə'tælədʒɪ] *n.* 冶金，冶金学，冶金术
metallurgist [mə'tælədʒɪst] *n.* 冶金家，冶金学者
Philosopher's Stone 点金石，魔法石
atomism ['ætəmɪzəm] *n.* 原子论，原子说
Ogdoad ['ɒgdəʊæd] *n.* 八元神
vague [veɪg] *adj.* 模糊的；（思想上）不清楚的；（表达或感知）含糊的；暧昧的
　　　　　　　　n. 模糊不定状态
alembic [ə'lembɪk] *n.* 蒸馏器
lapidary ['læpɪdərɪ] *adj.* 宝石的；简洁优雅的；刻在石上的；利落的
　　　　　　　　n. 宝石匠，玉石雕刻师
alkalis ['ælkəlɪs] *n.* 碱金属；碱（alkali 的名词复数）
affinity [ə'fɪnətɪ] *n.* 密切关系；类同；类似，近似
recurrent [rɪ'kʌrənt] *adj.* 复发的，复现的；周期性的，经常发生的；回归的；循环的
incidence ['ɪnsɪdəns] *n.* 发生率；影响范围；[数]关联，接合；[物]入射，入射角
plague [pleɪg] *n.* 瘟疫；灾害，折磨

      vt. 使染瘟疫；使痛苦，造成麻烦
blight [blaɪt] *n.* 凋萎病；坏因素；毁坏；衰退
      vt. 使凋萎；使颓丧；损害；妨害
      vi. 患枯萎病；枯萎，颓丧
Elixir of Life *n.* 仙丹
hybrid ['haɪbrɪd] *n.* 杂种；杂交生成的生物体；混合物；混合词
      adj. 混合的；杂种的
iatrochemistry [aɪˌætrə'kemɪstrɪ] *n.* 化学疗法
rigor ['rɪgə] *n.* 严格；严酷；严密
bias ['baɪəs] *n.* 偏见；偏爱，爱好；倾向；斜纹
      vt. 使倾向于；使有偏见；影响；加偏压于
      adj. 斜纹的；斜的，倾斜的；斜裁的
      adv. 偏斜地，倾斜地；对角地
gaseous ['gæsɪəs] *adj.* 气态的，似气体的；无实质的
conservation [ˌkɒnsə'veɪʃn] *n.* 保存；保护；避免浪费；对自然环境的保护
   conservation of mass　质量守恒定律
   conservation of energy　能量守恒定律
combustion [kəm'bʌstʃən] *n.* 燃烧，烧毁；氧化
overthrow [ˌəʊvə'θrəʊ] *vt.* 打倒，推翻；使屈服，征服；使瓦解；撞倒
      n. 推翻，打倒；打翻；倾倒
phlogiston [flɒ'dʒɪstən] *n.* （旧时人们认为存在于可燃物中的）燃素，热素
nomenclature [nə'menklətʃə] *n.* 系统命名法；命名（过程）；（某一学科的）术语；专门名称
metric ['metrɪk] *adj.* 米制的，公制的，十进制的；度量的；距离的
      n. 度量标准；[数学]度量；诗体，韵文，诗韵
archaic [ɑː'keɪɪk] *adj.* 古代的；过时的，陈旧的；古体的；古色古香的
synthesize ['sɪnθəsaɪz] *vt.* 综合；人工合成；（通过化学手段或生物过程）合成；（音响）合成
      vi. 合成；综合
culminate ['kʌlmɪneɪt] *vt.* & *vi.* 达到极点

Unit 2 Chemical Engineering

Chemical engineering is the key to many issues affecting our quality of life. Industry is increasingly focused on high value chemicals and products which deliver the right molecule to the right place at the right time. Chemical engineers conceive and design processes to produce, transform and transport materials—beginning with experimentation in the laboratory and followed by implementation of technologies in full-scale production.

Chemical engineering is a branch of engineering that uses principles of chemistry, physics, mathematics, and economics to efficiently use, produce, transform, and transport chemicals, materials, and energy. A chemical engineer designs large-scale processes that convert chemicals, raw materials, living cells, microorganisms, and energy into useful forms and products.

Chemical engineers are involved in many aspects of plant design and operation, including safety and hazard assessments, process design and analysis, control engineering, chemical reaction engineering, construction specification, and operating instructions.

Fig. 2-1 Chemical engineers design, construct and operate process plants.

2.1 Etymology

A 1996 *British Journal for the History of Science* article cites James F. Donnelly for mentioning an 1839 reference to chemical engineering in relation to the production of sulfuric acid. In the same paper however, George E. Davis, an English consultant, was credited for having coined the term. Davis also tried to find a Society of Chemical Engineering, but instead it was named the Society of Chemical Industry (1881), with Davis as its first Secretary. The *History of Science in*

United States: An Encyclopedia puts the use of the term around 1890. "Chemical engineering", describing the use of mechanical equipment in the chemical industry, became common vocabulary in England after 1850. By 1910, the profession, "chemical engineer", was already in common use in Britain and the United States.

Fig. 2-2   George E. Davis.

2.2   History of chemical engineering

Chemical engineering emerged upon the development of unit operations, a fundamental concept of the discipline of chemical engineering. Most authors agree that Davis invented the concept of unit operations if not substantially developed it. He gave a series of lectures on unit operations at the Manchester Technical School (later part of the University of Manchester) in 1887, considered to be one of the earliest such about chemical engineering. Three years before Davis' lectures, Henry Edward Armstrong taught a degree course in chemical engineering at the City and Guilds of London Institute. Armstrong's course failed simply because its graduates were not especially attractive to employers. Employers of the time would have rather hired chemists and mechanical engineers. Courses in chemical engineering offered by Massachusetts Institute of Technology (MIT) in the United States, Owens College in Manchester, England, and University College London suffered under similar circumstances.

Starting from 1888, Lewis M. Norton taught at MIT the first chemical engineering course in the United States. Norton's course was contemporaneous and essentially similar to Armstrong's course. Both courses, however, simply merged chemistry and engineering subjects along with product design. "Its practitioners had difficulty convincing engineers that they were engineers and chemists that they were not simply chemists." Unit operations was introduced into the course by William Hultz Walker in 1905. By the early 1920s, unit operations became an important aspect of chemical engineering at MIT and other US universities, as well as at Imperial College London. The American Institute of Chemical Engineers (AIChE), established in 1908, played a key role in making chemical engineering considered an independent science, and unit operations central to chemical engineering. For instance, it defined chemical engineering to be a "science of itself, the basis of which is ... unit operations" in a 1922 report; and with which principle, it had published a list of academic institutions which offered "satisfactory" chemical engineering courses. Meanwhile, promoting chemical engineering as a distinct science in Britain lead to the establishment of the Institution of Chemical Engineers (IChemE) in 1922. IChemE likewise helped make unit operations considered essential to the discipline.

2.3 New concepts and innovations

By the 1940s, it became clear that unit operations alone were insufficient in developing chemical reactors. While the predominance of unit operations in chemical engineering courses in Britain and the United States continued until the 1960s, transport phenomena started to experience greater focus. Along with other novel concepts, such process systems engineering (PSE), a "second paradigm" was defined. Transport phenomena gave an analytical approach to chemical engineering while PSE focused on its synthetic elements, such as control system and process design. Developments in chemical engineering before and after World War II were mainly incited by the petrochemical industry, however, advances in other fields were made as well. Advancements in biochemical engineering in the 1940s, for example, found application in the pharmaceutical industry, and allowed for the mass production of various antibiotics, including penicillin and streptomycin. Meanwhile, progress in polymer science in the 1950s paved way for the "age of plastics".

2.4 Safety and hazard developments

Concerns regarding the safety and environmental impact of large-scale chemical manufacturing facilities were also raised during this period. *Silent Spring*, published in 1962, alerted its readers to the harmful effects of DDT, a potent insecticide. The 1974 Flixborough disaster in the United Kingdom resulted in 28 deaths, as well as damage to a chemical plant and three nearby villages. The 1984 Bhopal disaster in India resulted in almost 4,000 deaths. These incidents, along with other incidents, affected the reputation of the trade as industrial safety and environmental protection were given more focus. In response, the IChemE required safety to be part of every degree course that it accredited after 1982. By the 1970s, legislation and monitoring agencies were instituted in various countries, such as France, Germany, and the United States.

2.5 Recent progress

Advancements in computer science found applications designing and managing plants, simplifying calculations and drawings that previously had to be done manually. The completion of the Human Genome Project is also seen as a major development, not only advancing chemical engineering but genetic engineering and genomics as well. Chemical engineering principles were used to produce DNA sequences in large quantities.

2.6 Concepts

Chemical engineering involves the application of several principles. Key concepts are presented below.

2.6.1 Plant design and construction

Chemical engineering design concerns the creation of plans, specification, and economic analyses for pilot plants, new plants or plant modifications. Design engineers often work in a consulting role, designing plants to meet clients' needs. Design is limited by a number of factors, including funding, government regulations and safety standards. These constraints dictate a plant's choice of process, materials and equipment.

Plant construction is coordinated by project engineers and project managers depending on the size of the investment. A chemical engineer may do the job of project engineer full-time or part of the time, which requires additional training and job skills or act as a consultant to the project group. In the USA the education of chemical engineering graduates from the Baccalaureate programs accredited by ABET do not usually stress project engineering education, which can be obtained by specialized training, as electives, or from graduate programs. Project engineering jobs are some of the largest employers for chemical engineers.

2.6.2 Process design and analysis

A unit operation is a physical step in an individual chemical engineering process. Unit operations (such as crystallization, filtration, drying and evaporation) are used to prepare reactants, purifying and separating its products, recycling unspent reactants, and controlling energy transfer in reactors. On the other hand, a unit process is the chemical equivalent of a unit operation. Along with unit operations, unit processes constitute a process operation. Unit processes (such as nitration and oxidation) involve the conversion of material by biochemical, thermochemical and other means. Chemical engineers responsible for these are called process engineers.

Process design requires the definition of equipment types and sizes as well as how they are connected together and the materials of construction. Details are often printed on a Process Flow Diagram which is used to control the capacity and reliability of a new or modified chemical factory.

Education for chemical engineers in the first college degree 3 or 4 years of study stresses the principles and practices of process design. The same skills are used in existing chemical plants to evaluate the efficiency and make recommendations for improvements.

2.6.3 Transport phenomena

Modeling and analysis of transport phenomena is essential for many industrial applications. Transport phenomena involve fluid dynamics, heat transfer and mass transfer, which are governed mainly by momentum transfer, energy transfer and transport of chemical species respectively. Models often involve separate considerations for macroscopic, microscopic and molecular level phenomena. Modeling of transport phenomena therefore requires an understanding of applied mathematics.

2.6.4 Applications and practice

Chemical engineers "develop economic ways of using materials and energy". Chemical engineers use chemistry and engineering to turn raw materials into usable products, such as medicine, petrochemicals and plastics on a large-scale, industrial setting. They are also involved in waste management and research. Both applied and research facets could make extensive use of computers.

Chemical engineers may be involved in industry or university research where they are tasked with designing and performing experiments to create better and safer methods for production, pollution control, and resource conservation. They may be involved in designing and constructing plants as a project engineer. Chemical engineers serving as project engineers use their knowledge in selecting optimal production methods and plant equipment to minimize costs and maximize safety and profitability. After plant construction, chemical engineering project managers may be involved in equipment upgrades, process changes, troubleshooting, and daily operations in either full-time or consulting roles.

New words

chemicals ['kemɪklz] *n.* 化学药品
conceive [kən'siːv] *vt. & vi.* 怀孕；构思；想象，设想；持有
             vi. 怀孕；设想；考虑
microorganism [ˌmaɪkrəʊ'ɔːɡənɪzəm] *n.* 微生物
hazard ['hæzəd] *vt.* 冒险；使遭受危险
          n. 危险；冒险的事；机会
etymology [ˌetɪ'mɒlədʒɪ] *n.* 词源学，词源说明
sulfuric [sʌl'fjuːrɪk] *adj.* [化]（正）硫的，含（六价）硫的
consultant [kən'sʌltənt] *n.* （受人咨询的）顾问；求教者，查阅者，咨询者
credit ['kredɪt] *n.* 信誉，信用；[金融]贷款；荣誉；学分
        vt. 相信，信任；归功于；[会]记入贷方；赞颂

substantially [səb'stænʃəlɪ] *adv.* 本质上，实质上；大体上；充分地；相当多地
contemporaneous [kən,tempə'reɪnɪəs] *adj.* 同时期的，同时代的
reactor [ri'æktə(r)] *n.* [化工]反应器；[核]反应堆；起反应的人；原子炉
paradigm ['pærədaɪm] *n.* 范例，样式，模范
incite [ɪn'saɪt] *vt.* 刺激；激励；煽动；促使
petrochemical [,petrəʊ'kemɪkl] *n.* 石油化学产品
　　　　　　　　　　　　 adj. 石油化学的；岩石化学的
pharmaceutical [,fɑːmə'suːtɪkl] *adj.* 制药的，配药的
　　　　　　　　　　　　　　 n. 药物
antibiotic [,æntibaɪ'ɒtɪk] *n.* 抗生素；抗菌素
　　　　　　　　　　　　 adj. 抗生的；抗菌的；抗菌作用的；抗生素的
penicillin [,penɪ'sɪlɪn] *n.* 青霉素，盘尼西林
streptomycin ['streptə'maɪsɪn] *n.* 链霉素
polymer ['pɒlɪmə] *n.* 多聚物；[高分子]聚合物
accredit [ə'kredɪt] *vt.* 归因于；委托，授权；相信，认可；鉴定合格，确认达标
institute ['ɪnstɪtjuːt] *vt.* 建立；制定；开始；着手
　　　　　　　　　　　 n. 协会；学会；学院；(教育、专业等)机构
genome ['dʒiːnəʊm] *n.* 基因组，染色体组
genomic [dʒiː'nəʊmɪk] *adj.* 染色体组的
genomics [dʒiː'nəʊmɪks] *n.* 基因组学
dictate [dɪk'teɪt] *vt.* 口述；命令，指示；使听写；控制，支配
　　　　　　　　　　 vi. 口述；命令；
　　　　　　　　　　 n. 命令；指示；指导原则
reactant [rɪ'æktənt] *n.* 反应物
nitration [naɪ'treɪʃən] *n.* 用硝酸处理，硝化，硝基置换
thermochemical [,θɜːməʊ'kemɪkəl] *adj.* 热化学的
dynamics [daɪ'næmɪks] *n.* 动力学，力学
momentum [mə'mentəm] *n.* [物]动量；动力
macroscopic [,mækrə'skɒpɪk] *adj.* 肉眼可见的；宏观的
microscopic [,maɪkrə'skɒpɪk] *adj.* 显微镜的；用显微镜可看见的；微小的；细微的
facet ['fæsɪt] *n.* 小平面；侧面，方面；(昆虫的)小眼面；[建]柱槽筋
　　　　　　　　 vt. 在……上琢面
troubleshooting ['trʌblʃuːtɪŋ] *n.* 发现并修理故障，解决纷争

Part 2

Basic Chemistry

Unit 3　Basic Concepts

3.1　Matter

In chemistry, matter is defined as anything that has rest mass and volume (it takes up space) and is made up of particles. The particles that make up matter have rest mass as well, not all particles have rest mass, such as the photon. Matter can be a pure chemical substance or a mixture of substances.

3.2　Atom

An atom is the basic unit of chemistry. It consists of a positively charged core (the atomic nucleus) which contains protons and neutrons, and which maintains a number of electrons to balance the positive charge in the nucleus. The atom is also the smallest entity that can be envisaged to retain some of the chemical properties of the element, such as electronegativity, ionization potential, preferred oxidation state(s), coordination number, and preferred types of bonds to form (e.g., metallic, ionic, covalent).

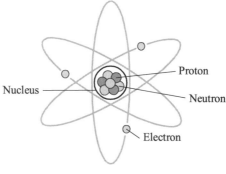

Fig. 3-1　Atom.

3.3　Element

The concept of chemical element is related to that of chemical substance. A chemical element is characterized by a particular number of protons in the nuclei of its atoms. This number is known as the atomic number of the element. For example, all atoms with 6 protons in their nuclei are atoms of the chemical element carbon, and all atoms with 92 protons in their nuclei are atoms of the element uranium. 94 different chemical elements or types of atoms based on the number of protons exist naturally. A further 18 have been recognised by IUPAC as existing artificially only. Although all the nuclei of all atoms belonging to one element will have the same number of protons, they may not necessarily have the same number of neutrons, such atoms are termed isotopes. In fact, several isotopes of an element may exist.

The most convenient presentation of the chemical elements is in the periodic table of the chemical elements, which groups elements by atomic number. Due to its ingenious arrangement, groups, or columns, and periods, or rows, of elements in the table either share several chemical properties, or follow a certain trend in characteristics such as atomic radius, electronegativity, etc. Lists of the elements by name, by symbol, and by atomic number are also available.

3.4 Compound

A compound is a substance with a particular ratio of atoms of particular chemical elements which determines its composition, and a particular organization which determines chemical properties. For example, water is a compound containing hydrogen and oxygen in the ratio of two to one, with the oxygen atom between the two hydrogen atoms, and an angle of 104.5° between them. Compounds are formed and interconverted by chemical reactions.

Fig. 3-2 Compound.

3.5 Substance

A chemical substance is a kind of matter with a definite composition and set of properties. Strictly speaking, a mixture of compounds, elements or compounds and elements is not a chemical substance, but it may be called a chemical. Most of the substances we encounter in our daily life are some kind of mixture, for example: air, alloys, biomass, etc.

Nomenclature of substances is a critical part of the language of chemistry. Generally, it refers to a system for naming chemical compounds. Earlier in the history of chemistry, substances were given name by their discoverer, which often led to some confusion and difficulty. However, today the IUPAC system of chemical nomenclature allows chemists to specify by name specific compounds amongst the vast variety of possible chemicals. The standard nomenclature of chemical substances is set by the International Union of Pure and Applied Chemistry (IUPAC). There are well-defined systems in place for naming chemical species. Organic compounds are named according to the organic nomenclature system. Inorganic compounds are named according to the inorganic nomenclature system. In addition, the Chemical Abstracts Service has devised a method to index chemical substance. In this scheme each chemical substance is identifiable by a number known as CAS registry number.

3.6 Molecule

A molecule is the smallest indivisible portion, besides an atom, of a pure chemical substance that has its unique set of chemical properties, that is, its potential to undergo a certain set of chemical reactions with other substances. Molecules can exist as electrically neutral units unlike ions. Molecules are typically a set of atoms bound together by covalent bonds, such that the structure is electrically neutral and all valence electrons are paired with other electrons either in bonds or in lone pairs.

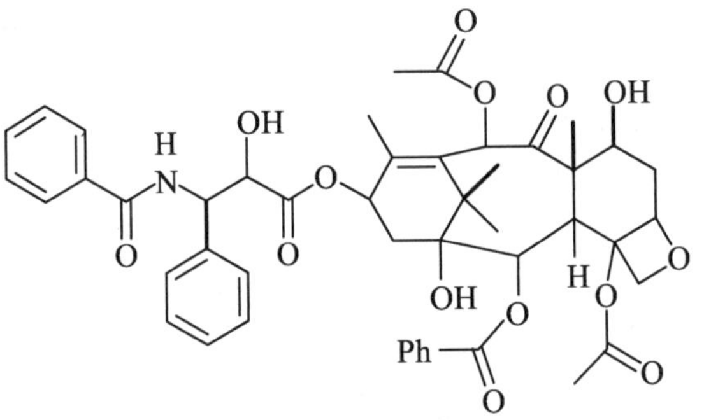

Fig. 3-3　A molecular structure depicts the bonds and relative positions of atoms in a molecule such as that in Paclitaxel shown here.

One of the main characteristic of a molecule is its geometry often called its structure. While the structure of diatomic, triatomic or tetra atomic molecules may be trivial, (linear, angular pyramidal etc.) the structure of polyatomic molecules, that are constituted of more than six atoms (of several elements) can be crucial for its chemical nature.

3.7 Mole

A mole is the amount of a substance that contains as many elementary entities (atoms, molecules or ions) as there are atoms in 0.012 kilogram (or 12 grams) of carbon-12, where the carbon-12 atoms are unbound, at rest and in their ground state. This number is known as the Avogadro constant, and is determined empirically. The currently accepted value is $6.02214179(30) \times 10^{23}$ mol^{-1} (2007 CODATA). The best way to understand the meaning of the term "mole" is to compare it to terms such as dozen. Just as one dozen is equal to 12, one mole is equal to $6.02214179(30) \times 10^{23}$. The term is used because it is much easier to say, for example, 1 mole of carbon atoms, than it is to say 6.02214179(30) × 10²³ carbon atoms. Likewise, we can describe the number of entities as a multiple or fraction of 1 mole, e.g. 2 mole or 0.5 moles. Mole is an absolute number (having no units) and can describe any type of elementary object, although the mole's use is usually limited to measurement of subatomic, atomic, and molecular structures.

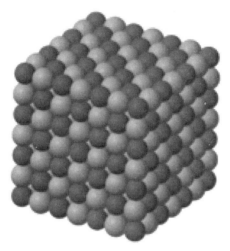

Fig. 3-4 The crystal lattice structure of potassium chloride (KCl), a salt which is formed due to the attraction of K⁺ cations and Cl⁻ anions. Note how the overall charge of the ionic compound is zero.

The number of moles of a substance in one liter of a solution is known as its molarity. Molarity is the common unit used to express the concentration of a solution in physical chemistry.

3.8 Ions and salts

An ion is a charged species, an atom or a molecule, that has lost or gained one or more electrons. Positively charged cations (e.g. sodium cation Na^+) and negatively charged anions (e.g. chloride Cl^-) can form a crystalline lattice of neutral salts (e.g. sodium chloride NaCl). Examples of polyatomic ions that do not split up during acid-base reactions are hydroxide (OH^-) and phosphate (PO_4^{3-}).

Ions in the gaseous phase is often known as plasma.

3.9 Acidity and basicity

A substance can often be classified as an acid or a base. This is often done on the basis of a particular kind of reaction, namely the exchange of protons between chemical compounds. However, an extension to this mode of classification was brewed up by the American chemist, Gilbert Newton Lewis; in this mode of classification the reaction is not limited to those occurring in an aqueous solution, thus is no longer limited to solutions in water. According to concept as per Lewis, the crucial things being exchanged are charges. There are several other ways in which a substance may be classified as an acid or a base, as is evident in the history of this concept.

Fig. 3-5 When hydrogen bromide (HBr), pictured, is dissolved in water, it forms the strong acid hydrobromic acid.

3.10 Phase

In addition to the specific chemical properties that distinguish different chemical classifications chemicals can exist in several phases. For the most part, the chemical classifications are independent of these bulk phase classifications; however, some more exotic phases are incompatible with certain chemical properties. A phase is a set of states of a chemical system that have similar bulk structural properties, over a range of conditions, such as pressure or temperature. Physical properties, such as density and refractive index tend to fall within values characteristic of the phase. The phase of matter is defined by the phase transition, which is when energy put into or taken out of the system goes into rearranging the structure of the system, instead of changing the bulk conditions.

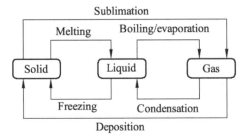

Fig. 3-6 Diagram showing relationships among the phases and the terms used to describe phase changes.

Sometimes the distinction between phases can be continuous instead of having a discrete boundary, in this case the matter is considered to be in a supercritical state. When three states meet based on the conditions, it is known as a triple point and since this is invariant, it is a convenient way to define a set of conditions.

The most familiar examples of phases are solids, liquids, and gases. Many substances exhibit multiple solid phases. For example, there are three phases of solid iron (alpha, gamma, and delta) that vary based on temperature and pressure. A principal difference between solid phases is the crystal structure, or arrangement, of the atoms. Another phase commonly encountered in the study of chemistry is the aqueous phase, which is the state of substances dissolved in aqueous solution (that is, in water). Less familiar phases include plasmas, Bose-Einstein condensates and fermionic condensates and the paramagnetic and ferromagnetic phases of magnetic materials. While most familiar phases deal with three-dimensional systems, it is also possible to define analogs in two-dimensional systems, which has received attention for its relevance to systems in biology.

3.11 Redox

It is a concept related to the ability of atoms of various substances to lose or gain electrons. Substances that have the ability to oxidize other substances are said to be oxidative and are known as oxidizing agents, oxidants or oxidizers. An oxidant removes electrons from another substance. Similarly, substances that have the ability to reduce other substances are said to be reductive and are known as reducing agents, reductants, or reducers. A reductant transfers electrons to another substance, and is thus oxidized itself. And because it "donates" electrons it is also called an electron donor. Oxidation and reduction properly refer to a change in oxidation number—the actual transfer of electrons may never occur. Thus, oxidation is better defined as an increase in oxidation number, and reduction as a decrease in oxidation number.

3.12 Bonding

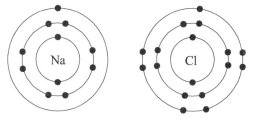

Fig. 3-7 An animation of the process of ionic bonding between sodium (Na) and chlorine (Cl) to form sodium chloride, or common table salt. Ionic bonding involves one atom taking valence electrons from another (as opposed to sharing, which occurs in covalent bonding).

Atoms sticking together in molecules or crystals are said to be bonded with one another. A chemical bond may be visualized as the multipole balance between the positive charges in the nuclei and the negative charges oscillating about them. More than simple attraction and repulsion, the energies and distributions characterize the availability of an electron to bond to another atom. These potentials create the interactions which hold atoms together in molecules or crystals. In many simple compounds, Valence Bond Theory, the Valence Shell Electron Pair Repulsion model (VSEPR), and the concept of oxidation number can be used to explain molecular structure and composition. Similarly, theories from classical physics can be used to predict many ionic structures. With more complicated compounds, such as metal complexes, valence bond theory fails and alternative approaches, primarily based on principles of quantum chemistry such as the molecular orbital theory, are necessary. See diagram on electronic orbitals.

3.13 Reaction

When a chemical substance is transformed as a result of its interaction with another or energy, a chemical reaction is said to have occurred. Chemical reaction is a therefore a concept related to the reaction of a substance when it comes in close contact with another, whether as a mixture or a solution; or its exposure to a some form of energy. It results in some energy exchange between the constituents of the reaction as well with the system environment which may be a designed vessels which are often laboratory glassware. Chemical reactions can result in the formation or dissociation of molecules, that is, molecules breaking apart to form two or more smaller molecules, or rearrangement of atoms within or across molecules. Chemical reactions usually involve the making or breaking of chemical bonds. Oxidation, reduction, dissociation, acid-base neutralization and molecular rearrangement are some of the commonly used kinds of chemical reactions.

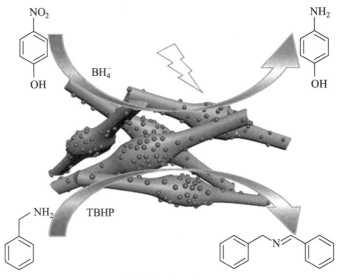

Fig. 3-8　Reaction.

A chemical reaction can be symbolically depicted through a chemical equation. While in a non-nuclear chemical reaction the number and kind of atoms on both sides of the equation are equal, for a nuclear reaction this holds true only for the nuclear particles, viz protons and neutrons.

The sequence of steps in which the reorganization of chemical bonds may be taking place in the course of a chemical reaction is called its mechanism. A chemical reaction can be envisioned to take place in a number of steps, each of which may have a different speed. Many reaction intermediates with variable stability can thus be envisaged during the course of a reaction. Reaction mechanisms are proposed to explain the kinetics and the relative product mix of a reaction. Many physical chemists specialize in exploring and proposing the mechanisms of various chemical reactions. Several empirical rules, like the Woodward-Hoffmann rules often come handy while proposing a mechanism for a chemical reaction.

According to the IUPAC gold book a chemical reaction is a process that results in the interconversion of chemical species. Accordingly, a chemical reaction may be an elementary reaction or a stepwise reaction. An additional caveat is made, in that this definition includes cases where the interconversion of conformers is experimentally observable. Such detectable chemical reactions normally involve sets of molecular entities as indicated by this definition, but it is often conceptually convenient to use the term also for changes involving single molecular entities (i.e. "microscopic chemical events").

3.14 Equilibrium

Although the concept of equilibrium is widely used across sciences, in the context of chemistry, it arises whenever a number of different states of the chemical composition are possible. For example, in a mixture of several chemical compounds that can react with one another, or when a substance can be present in more than one kind of phase. A system of chemical substances at equilibrium even though having an unchanging composition is most often not static; molecules of the substances continue to react with one another thus giving rise to a dynamic equilibrium. Thus the concept describes the state in which the parameters such as chemical composition remains unchanged over time. Chemicals present in biological systems are invariably not at equilibrium, rather they are far from equilibrium.

3.15 Energy

In the context of chemistry, energy is an attribute of a substance as a consequence of its atomic, molecular or aggregate structure. Since a chemical transformation is accompanied by a change in one or more of these kinds of structure, it is invariably accompanied by an increase or decrease of energy of the substances involved. Some energy is transferred between the surroundings and the reactants of the reaction in the form of heat or light; thus the products of a reaction may have more or less energy than the reactants. A reaction is said to be exergonic if the final state is lower on the energy scale than the initial state; in the case of endergonic reactions the situation is the reverse. A reaction is said to be exothermic if the reaction releases heat to the surroundings; in the case of endothermic reactions, the reaction absorbs heat from the surroundings.

Chemical reactions are invariably not possible unless the reactants surmount an energy barrier known as the activation energy. The speed of a chemical reaction (at given temperature T) is related to the activation energy E, by the Boltzmann's population factor $e^{-E/kT}$ — that is the probability of molecule to have energy greater than or equal to E at the given temperature T. This exponential dependence of a reaction rate on temperature is known as the Arrhenius equation. The activation energy necessary for a chemical reaction can be in the form of heat, light, electricity or mechanical force in the form of ultrasound.

A related concept free energy, which also incorporates entropy considerations, is a very useful means for predicting the feasibility of a reaction and determining the state of equilibrium of a chemical reaction, in chemical thermodynamics. A reaction is feasible only if the total change in the Gibbs free energy is negative, $\Delta G \leq 0$; if it is equal to zero the chemical reaction is said to be at equilibrium.

There exist only limited possible states of energy for electrons, atoms and molecules. These are determined by the rules of quantum mechanics, which require quantization of energy of a bound system. The atoms/molecules in a higher energy state are said to be excited. The molecules/atoms of substance in an excited energy state are often much more reactive; that is, more amenable to chemical reactions.

The phase of a substance is invariably determined by its energy and the energy of its surroundings. When the intermolecular forces of a substance are such that the energy of the surroundings is not sufficient to overcome them, it occurs in a more ordered phase like liquid or solid as is the case with water (H_2O); a liquid at room temperature because its molecules are bound by hydrogen bonds. Whereas hydrogen sulfide (H_2S) is a gas at room temperature and standard pressure, as its molecules are bound by weaker dipole-dipole interactions.

The transfer of energy from one chemical substance to another depends on the size of energy quanta emitted from one substance. However, heat energy is often transferred more easily from almost any substance to another because the phonons responsible for vibrational and rotational energy levels in a substance have much less energy than photons invoked for the electronic energy transfer. Thus, because vibrational and rotational energy levels are more closely spaced than electronic energy levels, heat is more easily transferred between substances relative to light or other forms of electronic energy. For example, ultraviolet electromagnetic radiation is not transferred with as much efficacy from one substance to another as thermal or electrical energy.

The existence of characteristic energy levels for different chemical substances is useful for their identification by the analysis of spectral lines. Different kinds of spectra are often used in chemical spectroscopy, e.g. IR, microwave, NMR, ESR, etc. Spectroscopy is also used to identify the composition of remote objects —like stars and distant galaxies —by analyzing their radiation spectra.

New words

rest mass 静止质量；静质量
nucleus ['njuːklɪəs] *n.* 中心，核心；（原子）核；起点，开始；[微]细胞核
entity ['entətɪ] *n.* 实体；实际存在物；本质
envisage [ɪn'vɪzɪdʒ] *vt.* 设想；想象；预想

electronegativity [elektrəʊnɪgə'tɪvɪtɪ] *n.* 阴电性，负电性，电负性；电阴性
ionization [,aɪənaɪ'zeɪʃn] *n.* 离子化，电离；电离化
coordination number 配位数
metallic [mə'tælɪk] *adj.* 金属般的；有金属味（或声音）的；金属制的；含金属的
ionic [aɪ'ɒnɪk] *adj.* 离子的
covalent [,kəʊ'veɪlənt] *adj.* 共有原子价的，共价的
isotope ['aɪsətəʊp] *n.* [化]同位素
ingenious [ɪn'dʒiːnɪəs] *adj.* 灵巧的；精巧的；设计独特的；有天才的，聪明的
radius ['reɪdɪəs] *n.* 半径（距离）；用半径度量的圆形面积；半径范围
valence ['veɪləns] *n.* （化合）价，原子价；（心理）效价
geometry [dʒɪ'ɒmətrɪ] *n.* 几何学；几何形状；几何图形；几何学著作
diatomic [,daɪə'tɒmɪk] *adj.* 二原子的，二氢氧基的，二价的
triatomic [,traɪə'tɒmɪk] *adj.* 由三个原子构成的，一分子中有三个原子的
tetra atomic [tetrə'tɒmɪk] *adj.* 四原子的
trivial ['trɪvɪəl] *adj.* 琐碎的，无价值的；平常的，平凡的；不重要的
linear ['lɪnɪə] *adj.* 直线的，线形的；长度的；<数>一次的，线性的
angular ['æŋgjələ(r)] *adj.* 有角的；用角测量的，用弧度测量的；生硬的，笨拙的
pyramidal ['pɪrəmɪdl] *adj.* 金字塔形的，锥体的
mole [məʊl] *n.* 摩尔
unbound ['ʌn'baʊnd] *adj.* 无束缚的
ground state 基态
constant ['kɒnstənt] *adj.* 不断的，持续的；永恒的，始终如一的；坚定；忠实的
　　　　　　n. [数]常数，常量；不变的事物；永恒值
empirically [ɪm'pɪrɪklɪ] *adv.* 以经验为主地
multiple ['mʌltɪpl] *adj.* 数量多的；多种多样的
　　　　　　n. 倍数
fraction ['frækʃn] *n.* [数]分数；一小部分，些微；不相连的一块，片段
molarity *n.* 体积摩尔（浓）度（每一公升溶液中溶质的摩尔数），体积克分子（浓）度；
　　　　　容模
ion ['aɪən] *n.* <物>离子
charged [tʃɑːdʒd] *adj.* 充满感情的；紧张的；带电的
chloride ['klɔːraɪd] *n.* 氯化物；〈口〉漂白粉
phosphate ['fɒsfeɪt] *n.* 磷酸盐；（含有少量磷酸的）汽水
plasma ['plæzmə] *n.* 血浆；原生质，细胞质；乳清；[物]等离子体
acidity [ə'sɪdətɪ] *n.* 酸性
basicity [be'sɪsətɪ] *n.* 碱度，碱性度
brew [bruː] *vt.* 酿造；泡，煮；策划，酿成
　　　　vi. 以酿造麦酒或啤酒为职业；被冲泡；即将发生
　　　　n. 酿造的饮料；啤酒

aqueous ['eɪkwɪəs] *adj.* 水的，水成的
phase [feɪz] *n.* 相，周相；阶段；[物理学]相位；方面，侧面
　　　　　　 vt. 分阶段实行；调整相位
　　　　　　 vi. 分阶段进行
bulk [bʌlk] *n.* （大）体积；大块，大量；大多数，大部分；主体
　　　　　　 vt. & vi. 变得越来越大（重要）；
　　　　　　 vi. 显得庞大；形成大块；堆积起来
　　　　　　 vt. 使凝聚成一团或形成一堆
　　　　　　 adj. 大批的，大量的；散装的
exotic [ɪg'zɒtɪk] *adj.* 异国的；外来的；异乎寻常的，奇异的；吸引人的
refractive [rɪ'fræktɪv] *adj.* 折射的
supercritical [ˌsju:pə'krɪtɪkəl] *adj.* 超临界的
invariant [ɪn'veərɪənt] *adj.* 无变化的，不变的
　　　　　　　　 n. 不变式，不变量
condensate ['kɒnd(ə)nseɪt] *n.* 冷凝物
paramagnetic [ˌpærəmæg'netɪk] *adj.* 顺磁性的
ferromagnetic [ˌferəʊmæg'netɪk] *adj.* 铁磁的，铁磁体的
analog ['ænəlɔ:g] *n.* 类似物，同源语；<电脑>模拟
　　　　　　 adj. （钟表）有长短针的；<电脑>模拟的
redox ['redɔks] *n.* 氧化还原作用
oxidative ['ɒksɪdeɪtɪv] *adj.* 氧化的，具有氧化特性的
oxidant *n.* 氧化剂
reductive [rɪ'dʌktɪv] *adj.* 减少的，还原的
　　　　　　　　 n. 还原剂
reductant [rɪ'dʌktənt] *n.* 还原剂
donor ['dəʊnə] *n.* 捐赠者，[医]供血者，输血者；[物]施主
bonding ['bɒndɪŋ] *n.* 黏结；连（搭、焊、胶、粘）接，结（耦、焊、接）合，压焊
amination ['æmɪneɪʃən] *n.* 氨基化，胺化，胺化作用
oscillate ['ɒsɪleɪt] *v.* 使振荡，使振动，使动摇
repulsion [rɪ'pʌlʃn] *n.* 厌恶；反感；<物>推斥；排斥力
dissociation [dɪˌsəʊsɪ'eɪʃn] *n.* 分离；离解；脱离关系；解体
neutralization [ˌnju:trəlaɪ'zeɪʃn] *n.* 中立化，中立状态，中和
viz [vɪz] *adv.* <正、书、尤英>即，就是
envisage [ɪn'vɪzɪdʒ] *vt.* 设想；想象；预想
kinetic [kɪ'netɪk] *adj.* 运动的，活跃的，能动的，有力的；[物]动力（学）的，运动的
caveat ['kævɪæt] *n.* 警告，附加说明
equilibrium [ˌi:kwɪ'lɪbrɪəm] *n.* 平衡，均势；平静
exergonic [ˌeksə'gɒnɪk] *adj.* 放出能的
endergonic [ˌendɜ:'gɒnɪk] *adj.* 吸能的

surmount [sə'maʊnt] *vt.* 战胜，克服；登上，攀登；居于……之上；顶上覆盖着
activation [ˌæktɪ'veɪʃn] *n.* 活化，激活，[化]活化作用；致活
exponential [ˌekspə'nenʃl] *n.* 指数；倡导者；演奏者；例子
     adj. 指数的，幂数的；越来越快的
ultrasound ['ʌltrəsaʊnd] *n.* 超声；超声波
amenable [ə'miːnəbl] *adj.* （对法律等）负责的；易控制的；经得起检验（考查）的；
    可用某种方式处理的
dipole ['daɪpəʊl] *n.* 双极子，偶极；振子；偶极子
phonon ['fəʊnɒn] *n.* 声子
vibrational [vaɪ'breɪʃənəl] *adj.* 振动的，摇摆的
rotational [rəʊ'teɪʃənl] *adj.* 转动的，轮流的
invoke [ɪn'vəʊk] *vt.* 乞灵，祈求；提出……以支持或证明；借助
ultraviolet [ˌʌltrə'vaɪələt] *adj.* 紫外的；紫外线的；产生紫外线的
     n. 紫外线辐射；紫外光
efficacy ['efɪkəsɪ] *n.* 功效；效力；效验；生产率
spectral ['spektrəl] *adj.* （似）鬼的；幽灵的；谱的；光谱的
spectroscopy [spek'trɒskəpɪ] *n.* [光] 光谱学
galaxy ['ɡæləksɪ] *n.* 星系；银河系；一群显赫的（出色的）人物

Unit 4　Subdisciplines

Chemistry is typically divided into several major sub-disciplines. There are also several main cross-disciplinary and more specialized fields of chemistry.

4.1　Analytical chemistry

Analytical chemistry is the analysis of material samples to gain an understanding of their chemical composition and structure. Analytical chemistry incorporates standardized experimental methods in chemistry. These methods may be used in all subdisciplines of chemistry, excluding purely theoretical chemistry.

Analytical chemistry studies and uses instruments and methods used to separate, identify, and quantify matter. In practice, separation, identification or quantification may constitute the entire analysis or be combined with another method. Separation isolates analytes. Qualitative analysis identifies analytes, while quantitative analysis determines the numerical amount or concentration.

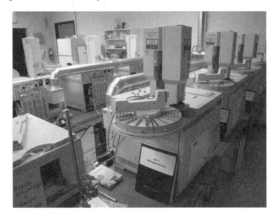

Fig. 4-1　Gas chromatography laboratory.

Analytical chemistry consists of classical, wet chemical methods and modern, instrumental methods. Classical qualitative methods use separations such as precipitation, extraction, and distillation. Identification may be based on differences in color, odor, melting point, boiling point, radioactivity or reactivity. Classical quantitative analysis uses mass or volume changes to quantify amount. Instrumental methods may be used to separate samples using chromatography, electrophoresis or field flow fractionation. Then qualitative and quantitative analysis can be performed, often with the same instrument and may use light interaction, heat interaction, electric fields or magnetic fields. Often the same instrument can separate, identify and quantify an analyte.

Analytical chemistry is also focused on improvements in experimental design, chemometrics, and the creation of new measurement tools. Analytical chemistry has broad applications to forensics, medicine, science and engineering.

Analytical chemistry has been important since the early days of chemistry, providing methods for determining which elements and chemicals are present in the object in question. During this period significant contributions to analytical chemistry include the development of systematic elemental analysis by Justus von Liebig and systematized organic analysis based on the specific reactions of functional groups.

The first instrumental analysis was flame emissive spectrometry developed by Robert Bunsen and Gustav Kirchhoff who discovered rubidium (Rb) and caesium (Cs) in 1860.

Most of the major developments in analytical chemistry take place after 1900. During this period instrumental analysis becomes progressively dominant in the field. In particular many of the basic spectroscopic and spectrometric techniques were discovered in the early 20th century and refined in the late 20th century.

The separation sciences follow a similar time line of development and also become increasingly transformed into high performance instruments. In the 1970s, many of these techniques began to be used together as hybrid techniques to achieve a complete characterization of samples.

Starting in approximately the 1970s into the present day, analytical chemistry has progressively become more inclusive of biological questions (bioanalytical chemistry), whereas it had previously been largely focused on inorganic or small organic molecules. Lasers have been increasingly used in chemistry as probes and even to initiate and influence a wide variety of reactions. The late 20th century also saw an expansion of the application of analytical chemistry from somewhat academic chemical questions to forensic, environmental, industrial and medical questions, such as in histology.

Modern analytical chemistry is dominated by instrumental analysis. Many analytical chemists focus on a single type of instrument. Academics tend to either focus on new applications and discoveries or on new methods of analysis. The discovery of a chemical present in blood that increases the risk of cancer would be a discovery that an analytical chemist might be involved in. An effort to develop a new method might involve the use of a tunable laser to increase the specificity and sensitivity of a spectrometric method. Many methods, once developed, are kept purposely static so that data can be compared over long periods of time. This is particularly true in industrial quality assurance (QA), forensic and environmental applications. Analytical chemistry plays an increasingly important role in the pharmaceutical industry where, aside from QA, it is used in discovery of new drug candidates and in clinical applications where understanding the interactions between the drug and the patient are critical.

4.2 Biochemistry

Biochemistry is the study of the chemicals, chemical reactions and chemical interactions that take place in living organisms. Biochemistry and organic chemistry are closely related, as in medicinal chemistry or neurochemistry. Biochemistry is also associated with molecular biology and genetics.

Biochemistry, sometimes called biological chemistry, is the study of chemical processes within and relating to living organisms. Biochemical processes give rise to the complexity of life.

A sub-discipline of both biology and chemistry, biochemistry can be divided in three fields: molecular genetics, protein science and metabolism. Over the last decades of the 20th century, biochemistry has through these three disciplines become successful at explaining living processes. Almost all areas of the life sciences are being uncovered and developed by biochemical methodology and research. Biochemistry focuses on understanding how biological molecules give rise to the processes that occur within living cells and between cells, which in turn relates greatly to the study and understanding of tissues, organs, and organism structure and function.

Fig. 4-2 Part of a series on biochemistry.

Biochemistry is closely related to molecular biology, the study of the molecular mechanisms by which genetic information encoded in DNA is able to result in the processes of life.

Much of biochemistry deals with the structures, functions and interactions of biological macromolecules, such as proteins, nucleic acids, carbohydrates and lipids, which provide the structure of cells and perform many of the functions associated with life. The chemistry of the cell also depends on the reactions of smaller molecules and ions. These can be inorganic, for example water and metal ions, or organic, for example the amino acids, which are used to synthesize proteins. The mechanisms by which cells harness energy from their environment via chemical reactions are known as metabolism. The findings of biochemistry are applied primarily in medicine, nutrition, and agriculture. In medicine, biochemists investigate the causes and cures of diseases. In nutrition, they study how to maintain health wellness and study the effects of nutritional deficiencies. In agriculture, biochemists investigate soil and fertilizers, and try to discover ways to improve crop cultivation, crop storage and pest control.

At its broadest definition, biochemistry can be seen as a study of the components and composition of living things and how they come together to become life, in this sense the history of biochemistry may therefore go back as far as the ancient Greeks. However, biochemistry as a specific scientific discipline has its beginning sometime in the 19th century, or a little earlier, depending on which aspect of biochemistry is being focused on. Some argued that the beginning of biochemistry may have been the discovery of the first enzyme, diastase (today called amylase), in 1833 by Anselme Payen, while others considered Eduard Buchner's first demonstration of a complex biochemical process alcoholic fermentation in cell-free extracts in 1897 to be the birth of biochemistry. Some might also point as its beginning to the influential 1842 work by Justus von Liebig, Animal chemistry, or, Organic chemistry in its applications to physiology and pathology, which presented a chemical theory of metabolism, or even earlier to the 18th century studies on fermentation and respiration by Antoine Lavoisier. Many other pioneers in the field who helped to uncover the layers of complexity of biochemistry have been proclaimed founders of modern biochemistry, for example Emil Fischer for his work on the chemistry of proteins, and F. Gowland Hopkins on enzymes and the dynamic nature of biochemistry.

The term "biochemistry" itself is derived from a combination of biology and chemistry. In 1877, Felix Hoppe-Seyler used the term (biochemie in German) as a synonym for physiological chemistry in the foreword to the first issue of Zeitschrift für Physiologische Chemie (Journal of Physiological Chemistry) where he argued for the setting up of institutes dedicated to this field of study. The German chemist Carl Neuberg however is often cited to have coined the word in 1903, while some credited it to Franz Hofmeister.

It was once generally believed that life and its materials had some essential property or substance (often referred to as the "vital principle") distinct from any found in non-living matter, and it was thought that only living beings could produce the molecules of life. Then, in 1828, Friedrich Wöhler published a paper on the synthesis of urea, proving that organic compounds can be created artificially. Since then, biochemistry has advanced, especially since the mid-20th century, with the development of new techniques such as chromatography, X-ray diffraction, dual polarisation interferometry, NMR spectroscopy, radioisotopic labeling, electron microscopy, and molecular dynamics simulations. These techniques allowed for the discovery and detailed analysis of many molecules and metabolic pathways of the cell, such as glycolysis and the Krebs cycle (citric acid cycle), and led to an understanding of biochemistry on a molecular level. Philip Randle is well known for his discovery in diabetes research, which is possibly the glucose-fatty acid cycle in 1963. He confirmed that fatty acids reduce oxidation of sugar by the muscle. High fat oxidation was responsible for the insulin resistance.

Another significant historic event in biochemistry is the discovery of the gene, and its role in the transfer of information in the cell. This part of biochemistry is often called molecular biology. In the 1950s, James D. Watson, Francis Crick, Rosalind Franklin, and Maurice Wilkins were instrumental in solving DNA structure and suggesting its relationship with genetic transfer of information. In 1958, George Beadle and Edward Tatum received the Nobel Prize for work in fungi showing that one gene produces one enzyme. In 1988, Colin Pitchfork was the first person convicted of murder with DNA evidence, which led to the growth of forensic science. More recently,

Andrew Z. Fire and Craig C. Mello received the 2006 Nobel Prize, for discovering the role of RNA interference (RNAi) in the silencing of gene expression.

4.3 Inorganic chemistry

Inorganic chemistry is the study of the properties and reactions of inorganic compounds. The distinction between organic and inorganic disciplines is not absolute and there is much overlap, most importantly in the sub-discipline of organometallic chemistry.

Inorganic chemistry deals with the synthesis and behavior of inorganic and organometallic compounds. This field covers all chemical compounds except the myriad organic compounds (carbon based compounds, usually containing C—H bonds), which are the subjects of organic chemistry. The distinction between the two disciplines is far from absolute, as there is much overlap in the subdiscipline of organometallic chemistry. It has applications in every aspect of the chemical industry, including catalysis, materials science, pigments, surfactants, coatings, medications, fuels, and agriculture.

Many inorganic compounds are ionic compounds, consisting of cations and anions joined by ionic bonding. Examples of salts (which are ionic compounds) are magnesium chloride $MgCl_2$, which consists of magnesium cations Mg^{2+} and chloride anions Cl^-; or sodium oxide Na_2O, which consists of sodium cations Na^+ and oxide anions O^{2-}. In any salt, the proportions of the ions are such that the electric charges cancel out, so that the bulk compound is electrically neutral. The ions are described by their oxidation state and their ease of formation can be inferred from the ionization potential (for cations) or from the electron affinity (anions) of the parent elements.

Important classes of inorganic compounds are the oxides, the carbonates, the sulfates, and the halides. Many inorganic compounds are characterized by high melting points. Inorganic salts typically are poor conductors in the solid state. Other important features include their high melting point and ease of crystallization. Where some salts (e.g., NaCl) are very soluble in water, others (e.g., SiO_2) are not.

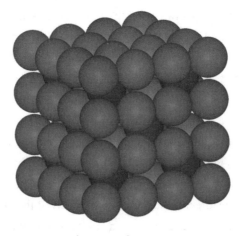

Fig. 4-3 The structure of the ionic framework in potassium oxide, K_2O.

The simplest inorganic reaction is double displacement when in mixing of two salts the ions are swapped without a change in oxidation state. In redox reactions one reactant, the oxidant, lowers its oxidation state and another reactant, the reductant, has its oxidation state increased. The net result is an exchange of electrons. Electron exchange can occur indirectly as well, e.g., in batteries, a key concept in electrochemistry.

When one reactant contains hydrogen atoms, a reaction can take place by exchanging protons in acid-base chemistry. In a more general definition, any chemical species capable of binding to electron pairs is called a Lewis acid; conversely any molecule that tends to donate an electron pair is referred to as a Lewis base. As a refinement of acid-base interactions, the HSAB theory takes into account polarizability and size of ions.

Inorganic compounds are found in nature as minerals. Soil may contain iron sulfide as pyrite or calcium sulfate as gypsum. Inorganic compounds are also found multitasking as biomolecules: as electrolytes (sodium chloride), in energy storage (ATP) or in construction (the polyphosphate backbone in DNA).

The first important man-made inorganic compound was ammonium nitrate for soil fertilization through the Haber process. Inorganic compounds are synthesized for use as catalysts such as vanadium (V) oxide and titanium (III) chloride, or as reagents in organic chemistry such as lithium aluminium hydride.

Subdivisions of inorganic chemistry are organometallic chemistry, cluster chemistry and bioinorganic chemistry. These fields are active areas of research in inorganic chemistry, aimed toward new catalysts, superconductors, and therapies.

4.4 Materials chemistry

Materials chemistry is the preparation, characterization, and understanding of substances with a useful function. The field is a new breadth of study in graduate programs, and it integrates elements from all classical areas of chemistry with a focus on fundamental issues that are unique to materials. Primary systems of study include the chemistry of condensed phases (solids, liquids, polymers) and interfaces between different phases.

The interdisciplinary field of materials science, also commonly termed materials science and engineering is the design and discovery of new materials, particularly solids. The intellectual origins of materials science stem from the Enlightenment, when researchers began to use analytical thinking from chemistry, physics, and engineering to understand ancient, phenomenological observations in metallurgy and mineralogy. Materials science still incorporates elements of physics, chemistry, and engineering. As such, the field was long considered by academic institutions as a sub-field of these related fields. Beginning in the 1940s, materials science began to be more widely recognized as a specific and distinct field of science and engineering, and major technical universities around the world created dedicated schools of the study, within either the Science or

Engineering schools, hence the naming.

Materials science is a syncretic discipline hybridizing metallurgy, ceramics, solid-state physics, and chemistry. It is the first example of a new academic discipline emerging by fusion rather than fission.

Many of the most pressing scientific problems humans currently face are due to the limits of the materials that are available and how they are used. Thus, breakthroughs in materials science are likely to affect the future of technology significantly.

Materials scientists emphasize understanding how the history of a material (its processing) influences its structure, and thus the material's properties and performance. The understanding of processing-structure-properties relationships is called the materials paradigm. This paradigm is used to advance understanding in a variety of research areas, including nanotechnology, biomaterials, and metallurgy. Materials science is also an important part of forensic engineering and failure analysis — investigating materials, products, structures or components which fail or do not function as intended, causing personal injury or damage to property. Such investigations are key to understanding, for example, the causes of various aviation accidents and incidents.

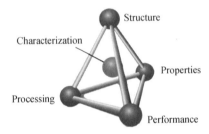

Fig. 4-4 The materials paradigm represented in the form of a tetrahedron.

4.5 Neurochemistry

Neurochemistry is the study of neurochemicals; including transmitters, peptides, proteins, lipids, sugars, and nucleic acids; their interactions, and the roles they play in forming, maintaining, and modifying the nervous system.

Neurochemistry is the study of neurochemicals, including neurotransmitters and other molecules such as psychopharmaceuticals and neuropeptides, that influence the function of neurons. This field within neuroscience examines how neurochemicals influence the operation of neurons, synapses, and neural networks. Neurochemists analyze the biochemistry and molecular biology of organic compounds in the nervous system, and their roles in such neural processes as cortical plasticity, neurogenesis, and neural differentiation.

In the 1950s, neurochemistry became a recognized scientific research discipline. The founding of neurochemistry as a discipline traces its origins to a series of "International Neurochemical Symposia", of which the first symposium volume published in 1954 was titled *Biochemistry of the Developing Nervous System*. These meetings led to the formation of the International Society for

Neurochemistry and the American Society for Neurochemistry. These early gatherings discussed the tentative nature of possible neurotransmitter substances such as acetylcholine, histamine, substance P, and serotonin. By 1972, ideas were more concrete. Neurochemicals such as norepinephrine, dopamine, and serotonin were classified as "putative neurotransmitters in certain neuronal tracts in the brain".

4.6 Nuclear chemistry

Nuclear chemistry is the study of how subatomic particles come together and make nuclei. Modern transmutation is a large component of nuclear chemistry, and the table of nuclides is an important result and tool for this field.

Nuclear chemistry is the subfield of chemistry dealing with radioactivity, nuclear processes, such as nuclear transmutation, and nuclear properties.

It is the chemistry of radioactive elements such as the actinides, radium and radon together with the chemistry associated with equipment (such as nuclear reactors) which are designed to perform nuclear processes. This includes the corrosion of surfaces and the behavior under conditions of both normal and abnormal operation (such as during an accident). An important area is the behavior of objects and materials after being placed into a nuclear waste storage or disposal site.

It includes the study of the chemical effects resulting from the absorption of radiation within living animals, plants, and other materials. The radiation chemistry controls much of radiation biology as radiation has an effect on living things at the molecular scale, to explain it another way the radiation alters the biochemicals within an organism, the alteration of the biomolecules then changes the chemistry which occurs within the organism, this change in chemistry then can lead to a biological outcome. As a result, nuclear chemistry greatly assists the understanding of medical treatments (such as cancer radiotherapy) and has enabled these treatments to improve.

It includes the study of the production and use of radioactive sources for a range of processes. These include radiotherapy in medical applications; the use of radioactive tracers within industry, science and the environment; and the use of radiation to modify materials such as polymers.

It also includes the study and use of nuclear processes in non-radioactive areas of human activity. For instance, nuclear magnetic resonance (NMR) spectroscopy is commonly used in synthetic organic chemistry and physical chemistry and for structural analysis in macromolecular chemistry.

4.7 Organic chemistry

Organic chemistry is a subdiscipline of chemistry that studies the structure, properties and reactions of organic compounds, which contain carbon in covalent bonding. Study of structure determines their chemical composition and formula. Study of properties includes physical and chemical properties, and evaluation of chemical reactivity to understand their behavior. The study of organic reactions includes the chemical synthesis of natural products, drugs, and polymers, and study of individual organic molecules in the laboratory and via theoretical (in silico) study.

The range of chemicals studied in organic chemistry includes hydrocarbons (compounds containing only carbon and hydrogen) as well as compounds based on carbon, but also containing other elements, especially oxygen, nitrogen, sulfur, phosphorus (included in many biochemicals) and the halogens. Organometallic chemistry is the study of compounds containing carbon-metal bonds.

In addition, contemporary research focuses on organic chemistry involving other organometallics including the lanthanides, but especially the transition metals zinc, copper, palladium, nickel, cobalt, titanium and chromium. Organic chemistry is the study of the structure, properties, composition, mechanisms, and reactions of organic compounds. An organic compound is defined as any compound based on a carbon skeleton.

Organic compounds form the basis of all earthly life and constitute the majority of known chemicals. The bonding patterns of carbon, with its valence of four — formal single, double, and triple bonds, plus structures with delocalized electrons — make the array of organic compounds structurally diverse, and their range of applications enormous. They form the basis of, or are constituents of, many commercial products including pharmaceuticals, petrochemicals and agrichemicals, and products made from them including lubricants, solvents, plastics, fuels and explosives. The study of organic chemistry overlaps organometallic chemistry and biochemistry, but also with medicinal chemistry, polymer chemistry, and materials science.

Before the nineteenth century, chemists generally believed that compounds obtained from living organisms were endowed with a vital force that distinguished them from inorganic compounds. According to the concept of vitalism (vital force theory), organic matter was endowed with a "vital force". During the first half of the nineteenth century, some of the first systematic studies of organic compounds were reported. Around 1816, Michel Chevreul started a study of soaps made from various fats and alkalis. He separated the different acids that, in combination with the alkali, produced the soap. Since these were all individual compounds, he demonstrated that it was possible to make a chemical change in various fats (which traditionally come from organic sources), producing new compounds, without "vital force". In 1828 Friedrich Wöhler produced the organic chemical urea (carbamide), a constituent of urine, from inorganic starting materials (the salts potassium cyanate and ammonium sulfate), in what is now called the Wöhler synthesis. Although Wöhler himself was cautious about claiming he had disproved vitalism, this was the first time a substance thought to be organic was synthesized in the laboratory without biological (organic) starting materials. The event is now generally accepted as indeed disproving the doctrine of vitalism.

Fig. 4-5 Friedrich Wöhler

In 1856 William Henry Perkin, while trying to manufacture quinine accidentally produced the organic dye now known as Perkin's mauve. His discovery, made widely known through its financial success, greatly increased interest in organic chemistry.

A crucial breakthrough for organic chemistry was the concept of chemical structure, developed independently in 1858 by both Friedrich August Kekulé and Archibald Scott Couper. Both researchers suggested that tetravalent carbon atoms could link to each other to form a carbon lattice, and that the detailed patterns of atomic bonding could be discerned by skillful interpretations of appropriate chemical reactions.

The era of the pharmaceutical industry began in the last decade of the 19th century, when the manufacturing of acetylsalicylic acid—more commonly referred to as aspirin—in Germany was started by Bayer. By 1910 Paul Ehrlich and his laboratory group began developing arsenic-based arsphenamine (Salvarsan), as the first effective medicinal treatment of syphilis, and thereby initiated the medical practice of chemotherapy. Ehrlich popularized the concepts of "magic bullet" drugs and of systematically improving drug therapies. His laboratory made decisive contributions to developing antiserum for diphtheria and standardizing therapeutic serums.

Early examples of organic reactions and applications were often found because of a combination of luck and preparation for unexpected observations. The latter half of the 19th century however witnessed systematic studies of organic compounds. The development of synthetic indigo is illustrative. The production of indigo from plant sources dropped from 19,000 tons in 1897 to 1,000 tons by 1914, thanks to the synthetic methods developed by Adolf von Baeyer. In 2002, 17,000 tons of synthetic indigo were produced from petrochemicals.

In the early part of the 20th century, polymers and enzymes were shown to be large organic molecules, and petroleum was shown to be of biological origin.

The multiple-step synthesis of complex organic compounds is called total synthesis. Total synthesis of complex natural compounds increased in complexity to glucose and terpineol. For example, cholesterol-related compounds have opened ways to synthesize complex human hormones and their modified derivatives. Since the start of the 20th century, complexity of total syntheses has been increased to include molecules of high complexity such as lysergic acid and vitamin B_{12}.

The discovery of petroleum and the development of the petrochemical industry spurred the development of organic chemistry. Converting individual petroleum compounds into different types of compounds by various chemical processes led to organic reactions enabling a broad range of industrial and commercial products including, among (many) others: plastics, synthetic rubber, organic adhesives, and various property-modifying petroleum additives and catalysts.

The majority of chemical compounds occurring in biological organisms are in fact carbon compounds, so the association between organic chemistry and biochemistry is so close that biochemistry might be regarded as in essence a branch of organic chemistry. Although the history of biochemistry might be taken to span some four centuries, fundamental understanding of the field only began to develop in the late 19th century and the actual term biochemistry was coined around the start of 20th century. Research in the field increased throughout the 20th century, without any indication of slackening in the rate of increase, as may be verified by inspection of abstraction and indexing services such as BIOSIS Previews and Biological Abstracts, which began in the 1920s as a single annual volume, but has grown so drastically that by the end of the 20th century it was only available to the everyday user as an online electronic database.

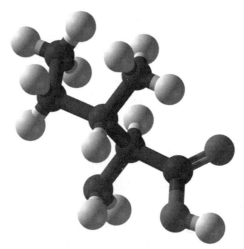

Fig. 4-6 The L-isoleucine molecule, $C_6H_{13}NO_2$, showing features typical of organic compounds. Carbon atoms are in black, hydrogens gray, oxygens red, and nitrogen blue.

4.8 Physical chemistry

Physical chemistry is the study of the physical and fundamental basis of chemical systems and processes. In particular, the energetics and dynamics of such systems and processes are of interest to physical chemists. Important areas of study include chemical thermodynamics, chemical kinetics, electrochemistry, statistical mechanics, and spectroscopy. Physical chemistry has large overlap with molecular physics. Physical chemistry involves the use of infinitesimal calculus in deriving equations. It is usually associated with quantum chemistry and theoretical chemistry. Physical chemistry is a distinct discipline from chemical physics.

Physical chemistry is the study of macroscopic, atomic, subatomic, and particulate phenomena in chemical systems in terms of the principles, practices, and concepts of physics such as motion, energy, force, time, thermodynamics, quantum chemistry, statistical mechanics, analytical dynamics and chemical equilibrium.

Physical chemistry, in contrast to chemical physics, is predominantly (but not always) a macroscopic or supra-molecular science, as the majority of the principles on which it was founded relate to the bulk rather than the molecular/atomic structure alone (for example, chemical equilibrium and colloids).

Some of the relationships that physical chemistry strives to resolve include the effects of:

1. Intermolecular forces that act upon the physical properties of materials (plasticity, tensile strength, surface tension in liquids).

2. Reaction kinetics on the rate of a reaction.

3. The identity of ions and the electrical conductivity of materials.

4. Surface science and electrochemistry of cell membranes.

5. Interaction of one body with another in terms of quantities of heat and work called thermodynamics.

6. Transfer of heat between a chemical system and its surroundings during change of phase or chemical reaction taking place called thermochemistry.

7. Study of colligative properties of number of species present in solution.

8. Number of phases, number of components and degree of freedom (or variance) can be correlated with one another with the help of phase rule.

9. Reactions of electrochemical cells.

4.9 Theoretical chemistry

Theoretical chemistry is the study of chemistry via fundamental theoretical reasoning (usually within mathematics or physics). In particular the application of quantum mechanics to chemistry is called quantum chemistry. Since the end of the World War II, the development of computers has allowed a systematic development of computational chemistry, which is the art of developing and applying computer programs for solving chemical problems. Theoretical chemistry has large overlap with (theoretical and experimental) condensed matter physics and molecular physics.

Theoretical chemistry is a branch of chemistry, which develops theoretical generalizations that are part of the theoretical arsenal of modern chemistry, for example, the concept of chemical bonding, chemical reaction, valence, the surface of potential energy, molecular orbitals, orbital interactions, molecule activation etc.

Theoretical chemistry unites principles and concepts common to all branches of chemistry. Within the framework of theoretical chemistry, there is a systematization of chemical laws, principles and rules, their refinement and detailing, the construction of a hierarchy. The central place in theoretical chemistry is occupied by the doctrine of the interconnection of the structure and properties of molecular systems. It uses mathematical and physical methods to explain the structures and dynamics of chemical systems and to correlate, understand, and predict their thermodynamic and kinetic properties. In the most general sense, it is explanation of chemical phenomena by methods of theoretical physics. In contrast to theoretical physics, in connection with the high complexity of chemical systems, theoretical chemistry, in addition to approximate mathematical methods, often uses semi-empirical and empirical methods.

Fig. 4-7　J. van't Hoff (1852–1911), the first winner of the Nobel Prize in Chemistry, is widely considered one of the most brilliant theoretical chemists in history.

In recent years, it has consisted primarily of quantum chemistry, i.e., the application of quantum mechanics to problems in chemistry. Other major components include molecular dynamics, statistical thermodynamics and theories of electrolyte solutions, reaction networks, polymerization, catalysis, molecular magnetism and spectroscopy.

Modern theoretical chemistry may be roughly divided into the study of chemical structure and the study of chemical dynamics. The former includes studies of: electronic structure, potential energy surfaces, and force fields; vibrational-rotational motion; equilibrium properties of condensed-phase systems and macro-molecules. Chemical dynamics includes: bimolecular kinetics and the collision theory of reactions and energy transfer; unimolecular rate theory and metastable states; condensed-phase and macromolecular aspects of dynamics.

Other fields include agrochemistry, astrochemistry, atmospheric chemistry, chemical engineering, chemical biology, chemo-informatics, electrochemistry, environmental chemistry, femtochemistry, flavor chemistry, flow chemistry, geochemistry, green chemistry, histochemistry, history of chemistry, hydrogenation chemistry, immunochemistry, marine chemistry, materials science, mathematical chemistry, mechanochemistry, medicinal chemistry, molecular biology, molecular mechanics, nanotechnology, natural product chemistry, oenology, neurochemistry, organometallic chemistry, petrochemistry, pharmacology, photochemistry, physical organic chemistry, phytochemistry, polymer chemistry, radiochemistry, solid-state chemistry, sonochemistry, supramolecular chemistry, surface chemistry, synthetic chemistry, thermochemistry, and many others.

New words

subdiscipline ['sʌb'dɪsɪplɪn] *n.* （学科的）分支，分科
cross-disciplinary [,krɒs dɪsə'plɪnərɪ] *adj.* 交叉训练的；跨学科的；交叉学科的
analyte [ænə'lɪt] *n.* （被）分析物；分解物
qualitative ['kwɒlɪtətɪv] *adj.* 定性的，定质的；性质上的；质量的
quantitative ['kwɒntɪtətɪv] *adj.* 定量的；数量（上）的
precipitation [prɪ,sɪpɪ'teɪʃn] *n.* 匆促；沉淀；（雨等）降落；某地区降雨等的量
distillation [,dɪstɪ'leɪʃn] *n.* （各种释义的）蒸馏（过程）；蒸馏物；蒸流；升华
odor ['əʊdə] *n.* 气味，名声
chromatography [,krəʊmə'tɒgrəfɪ] *n.* 套色版；层析法
electrophoresis [ɪ,lektrəʊfə'riːsɪs] *n.* 电泳疗法
fractionation [,frækʃən'eɪʃən] *n.* 分别，分馏法
chemometrics ['kemәʊ'metrɪks] *n.* 化学计量学，化学统计学
forensics [fə'rensɪks] *n.* 辩论练习，辩论术
genetics [dʒə'netɪks] *n.* 遗传学
protein ['prəʊtiːn] *n.* [化]朊，蛋白（质）
　　　　　　　　　　adj. 蛋白质的
metabolism [mə'tæbəlɪzəm] *n.* 新陈代谢；代谢作用
tissue ['tɪʃuː] *n.* 薄纸，棉纸；[生]组织；一套
macromolecule [,mækrə'mɒləkjuːl] *n.* 巨大分子，高分子
carbohydrate [,kɑːbəʊ'haɪdreɪt] *n.* 碳水化合物，糖类；淀粉质或糖类食物
lipid ['lɪpɪd] *n.* <生化>脂质
amino [ə'miːnəʊ] *adj.* <化>氨基的
harness ['hɑːnɪs] *n.* 马具，挽具；背带；系带
　　　　　　　　　vt. 利用；给（马等）套轭具；控制
organometallic [,ɔːgənəʊmə'tælɪk] *adj.* 有机金属的
myriad ['mɪrɪəd] *adj.* 无数的；多种的，各式各样的
　　　　　　　　　n. 无数，极大数量
catalysis [kə'tæləsɪs] *n.* 催化作用

pigment ['pɪgmənt] *n.* 颜料，色料；[生]色素
　　　　　　　　　vt. 给……着色
　　　　　　　　　vi. 呈现颜色
surfactant [sɜː'fæktənt] *n. & adj.* 表面活性剂（的）
coating ['kəʊtɪŋ] *n.* 涂层，覆盖层；（食品上的）面衣，糖衣，涂料；外衣布料，上衣料，
　　　　　　　细呢，花呢
　　　　　　v. 给……穿上外衣；覆盖
medication [ˌmedɪ'keɪʃn] *n.* 药物，药剂；药物治疗；药物处理；加入药物
polymer ['pɒlɪmə(r)] *n.* 多聚物；[高分子] 聚合物
interface ['ɪntəfeɪs] *n.* 界面；<计>接口；交界面
　　　　　　　　v. （使通过界面或接口）接合，连接；[计算机]使联系
　　　　　　　　vi. 相互作用（或影响）；交流，交谈
phenomenology [fɪˌnɒmɪ'nɒlədʒɪ] *n.* 现象学；唯象论
metallurgy [mə'tælədʒɪ] *n.* 冶金，冶金学，冶金术
mineralogy [ˌmɪnə'rælədʒɪ] *n.* 矿物学
syncretic [sɪnk'retɪk] *adj.* 融合的；汇合的；类并的
hybridize ['haɪbrɪdaɪz] *v.* （使）杂交
ceramics [sɪ'ræmɪks] *n.* 制陶术；陶器制造；陶瓷工艺；陶瓷制品（ceramic 的名词复数），
　　　　　　陶瓷器；制陶艺术；陶瓷装潢艺术
fusion ['fjuːʒn] *n.* 融合；熔解，熔化；融合物；[物]核聚变
fission ['fɪʃn] *n.* <物>（原子的）分裂，裂变；<生>分裂生殖
nanotechnology [ˌnænəʊtek'nɒlədʒɪ] *n.* 纳米技术，毫微技术；纤技术
aviation [ˌeɪvi'eɪʃn] *n.* 航空；飞行术；航空学；飞机制造业；[集合词]飞机
transmitter [træns'mɪtə] *n.* 传送者；传达者；发射机；发报机
peptide ['peptaɪd] *n.* <生化>肽，缩氨酸
psychopharmaceutical [saɪkəʊfɑːmə'sjuːtɪkəl] *adj. & n.* 精神（病）药物（的）
neuropeptide [njʊərə'peptaɪd] *n.* 神经肽
neuron ['njʊərɒn] *n.* 神经元；神经细胞
synapse ['saɪnæps] *n.* <生>（神经元的）突触
cortical ['kɔːtɪkl] *adj.* 皮层的，皮质的，有关脑皮层的
plasticity [plæ'stɪsətɪ] *n.* 黏性；成形性；柔软性；<生>可塑性（指生物体对环境的适应性）
neurogenesis [njʊərə'dʒenɪsɪs] *n.* 神经形成
differentiation [ˌdɪfəˌrenʃɪ'eɪʃn] *n.* 区别，分化；分异；衍进；求导数
symposium [sɪm'pəʊzɪəm] *n.* 专题讨论会，座谈会，学术报告会；专题论文集
　　　　　　　　复数形式：symposia or symposiums
acetylcholine [ˌæsɪtɪl'kɒliːn] *n.* 乙酰胆碱
histamine ['hɪstəmiːn] *n.* 组胺
serotonin [ˌserə'təʊnɪn] *n.* [医]5-羟色胺
norepinephrine [ˌnɔːrepɪ'nefrɪn] *n.* 降肾上腺素，去甲肾上腺素

dopamine ['dəʊpəmiːn] *n.* <生化>多巴胺
putative ['pjuːtətɪv] *adj.* 一般认定的，推定的，假定存在的
transmutation [ˌtrænsmjuːˈteɪʃn] *n.* 变形，变化
nuclide ['nuːklaɪd] *n.* 核素
radioactivity [ˌreɪdɪəʊækˈtɪvətɪ] *n.* 放射（性）；辐射能
actinide ['æktɪnaɪd] *n.* 锕类
radium ['reɪdɪəm] *n.* <化>镭（88号元素符号Ra）
radon ['reɪdɒn] *n.* <化>氡（元素符号Rn）
corrosion [kəˈrəʊʒn] *n.* 腐蚀，侵蚀，锈蚀；受腐蚀的部位；衰败
absorption [əbˈsɔːpʃn] *n.* 吸收；专注；合并
radiation [ˌreɪdɪˈeɪʃn] *n.* 辐射；放射物；辐射状；分散
radiotherapy [ˌreɪdɪəʊˈθerəpɪ] *n.* 放射疗法
resonance ['rezənəns] *n.* 共鸣；反响；共振
silico [sɪˈlɪkəʊ] *abbr.* silicosis *n.* 硅肺
hydrocarbon [ˌhaɪdrəˈkɑːbən] *n.* <化>碳氢化合物，烃
nitrogen ['naɪtrədʒən] *n.* [化]氮，氮气
sulfur ['sʌlfə] *n.* 硫黄，硫黄；硫黄色
　　　　 vt. 用硫黄处理
phosphorus ['fɒsfərəs] *n.* [化]磷；磷光体
halogen ['hælədʒən] *n.* 卤素
organometallic [ˌɔːɡənəʊməˈtælɪk] *adj.* 有机金属的
alkali ['ælkəlaɪ] *n.* 碱
lithium ['lɪθɪəm] *n.* <化>锂
sodium ['səʊdɪəm] *n.* <化>钠
potassium [pəˈtæsɪəm] *n.* <化>钾
alkaline ['ælkəlaɪn] *adj.* 碱性的，碱的；含碱的
　　　　　 n. 碱度，碱性
magnesium [mæɡˈniːzɪəm] *n.* [化]镁
metalloid ['metlɔɪd] *n.* 类金属
　　　　　 adj. 类似金属性的
boron ['bɔːrɒn] *n.* <化>硼
aluminium [ˌæljəˈmɪnɪəm] *n.* 铝
　　　　　　 adj. 铝的
tin [tɪn] *n.* 锡；罐头盒；马口铁 镀锡薄钢板
　　 adj. 锡制的；假冒的；无价值的；蹩脚的
　　 vt. 镀锡，包锡；给……包马口铁；包白铁
lanthanide ['lænθənaɪd] *n.* 镧系元素（原子序数58至81间的稀土元素）
palladium [pəˈleɪdɪəm] *n.* 钯
delocalize [diːˈləʊkəlaɪz] *vt.* 使离开原位

array [ə'reɪ] *n.* 队列，阵列；数组；一大批；衣服
　　　　vt. 排列；部署兵力；打扮，装饰
pharmaceuticals [ˌfɑːməˈsjuːtɪklz] *n.* 医药品；药物（pharmaceutical 的名词复数）
petrochemicals [ˌpetrəʊˈkemɪkəlz] *n.* 石油化学产品（petrochemical 的名词复数）
agrichemical 农用化学品，用农产品制得的化学品
lubricant [ˈluːbrɪkənt] *n.* 润滑剂，润滑油；能减少摩擦的东西
　　　　adj. 润滑的
solvent [ˈsɒlvənt] *n.* [化]溶剂，溶媒；解释，说明；解决方法；使瓦解的东西
　　　　adj. 有溶解能力的，可溶解的；有清还债务能力的；起瓦解作用的
energetics [ˌenəˈdʒetɪks] *n.* 热力学，动能学，力能学
dynamics [daɪˈnæmɪks] *n.* 动力学，力学；动态动力（dynamic 的名词复数）
thermodynamics [ˌθɜːməʊdaɪˈnæmɪks] *n.* 热力学
kinetics [kɪˈnetɪks] *n.* 动力学
electrochemistry [ɪˌlektrəʊˈkemɪstrɪ] *n.* 电化学
mechanics [mɪˈkænɪks] *n.* 力学；机械学
infinitesimal [ˌɪnfɪnɪˈtesɪml] *adj.* 极微小的
　　　　adv. 无限小地
　　　　n. 无限小；极微量；极小量
calculus [ˈkælkjələs] *n.* [医]结石，石，积石，牙垢；[数]运算，演算，微积分（学）
quantum [ˈkwɒntəm] *n.* [物]量子；定量，总量
particulate [pɑːˈtɪkjələt] *adj.* 微粒的，粒子的
　　　　n. 微粒，粒子
equilibrium [ˌiːkwɪˈlɪbrɪəm] *n.* 平衡，均势；平静
colloid [ˈkɒlɔɪd] *n. & adj.* 胶体（的）
membrane [ˈmembreɪn] *n.* （动物或植物体内的）薄膜；隔膜；（防水、防风用的）膜状物
colligative [ˈkɒlɪˌɡeɪtɪv] *adj.* （物质的物理性质）依数的
arsenal [ˈɑːsənl] *n.* 兵工厂，军火库
orbital [ˈɔːbɪtl] *adj.* 轨道的；眼窝的
doctrine [ˈdɒktrɪn] *n.* 教条，教义；法律原则；声明
approximate [əˈprɒksɪmət] *adj.* 约莫的，大概的；极相似的；[植]相近但不连接的
　　　　vi. 接近于；近似于
　　　　vt. 靠近；使接近；使结合
electrolyte [ɪˈlektrəlaɪt] *n.* <化>电解液，电解质
polymerization [ˌpɒlɪməraɪˈzeɪʃn] *n.* 聚合；多项式
magnetism [ˈmæɡnətɪzəm] *n.* 磁性，磁力；磁学；吸引力；催眠术
metastable [ˈmetəˈsteɪbəl] *adj.* 亚稳的
femtochemistry [femtʊˈkemɪstrɪ] *n.* [医]飞秒化学

Unit 5 Pharmaceutical engineering

Pharmaceutical engineering is a field that deals with the process of creating manufacturing plants and the pharmaceutical products that these plants generate. A pharmaceutical engineer helps to produce regulatory guidelines regarding the production of medical drugs as well. The field of biotechnology typically requires individuals to complete a minimum of four years of college-level training and to get practical experience with handling the pharmaceutical production process.

Fig. 5-1 Manufacturing plants are an important part of pharmaceutical engineering.

An important aspect of the pharmaceutical engineering industry includes the specialty areas of upholding regulatory standards and facilitating the delivery of pharmaceutical products. A pharmaceutical engineer makes sure that personal and environmental safety standards are being maintained when pharmaceuticals are being produced. In addition, a biotechnologist is responsible for labeling and packaging pharmaceuticals after validating the integrity of the end products.

Fig. 5-2 Biotechnology is an important component of pharmaceutical engineering.

Another major aspect of the biotechnology industry is the pharmaceutical facility design process. Pharmaceutical engineers essentially construct both pharmaceutical manufacturing plants and research facilities while taking into consideration the design of process equipment as well as "cleanrooms", important utilities and water systems. Cleanrooms are enclosed areas that have a low number of environmental pollutants, such as dust and other contaminants.

Fig. 5-3　Pharmaceutical engineers often design and construct pharmaceutical manufacturing plants.

The development of high-quality drugs is a chief part of the pharmaceutical engineering industry as well. Pharmaceutical engineers use their knowledge of important drug attributes along with chemical processes and scientific computing procedures to put together drugs that target various health conditions. Biotechnologists essentially facilitate the conversion of biological and chemical materials into drugs that humans can use, and also perform quality assurance tests to ensure that these drugs fulfill their intended purposes.

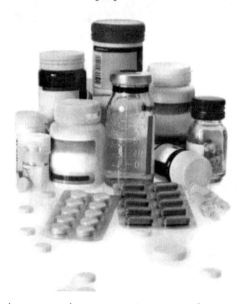

Fig. 5-4　A pharmaceutical engineer works to create pharmaceutical products.

People who wish to work in the biotechnology industry have a multitude of career options and must possess several skills. Positions for pharmaceutical engineers are advertised as positions in biochemistry, as bioprocesses engineering jobs, or as chemical engineering opportunities. Pharmaceutical engineering industry jobs can be found at sites such as government agencies, colleges, pharmaceutical companies, and even national labs. All career options in the pharmaceutical engineering industry demand that individuals are strong leaders and possess solid organizational, communication, and interpersonal skills.

Entry into the pharmaceutical engineering industry requires a minimum of a four-year bachelor's degree in a science or engineering area. Employers, however, typically look for pharmaceutical engineers who have a two-year master's degree or a four- to five-year doctoral degree in the industry. Graduate degrees typically are required for a pharmaceutical engineer who wishes to work in research. A pharmaceutical engineer also can pursue voluntary industry certification through the International Society for Pharmaceutical Engineering.

Fig. 5-5 Many pharmaceutical engineers have an extensive background in chemistry.

5.1 History

Humans have a long history of using derivatives of natural resources, such as plants and medication. However, it was not until the late 19th century when the technological advancements of chemical companies were combined with medical research that scientists began to manipulate and engineer new medications, drug delivery techniques, and methods of mass production.

5.1.1 Synthesizing new medications

One of the first prominent examples of an engineered, synthetic medication was made by Paul Erlich. Erlich had found that Atoxyl, an arsenic-containing compound which is harmful to humans, was very effective at killing Treponema pallidum, the bacteria which causes Syphilis. He hypothesized that if the structure of Atoxyl was altered, a "magic bullet" could potentially be identified which would kill the parasitic bacteria without having any adverse effects on human health. He developed many compounds stemming from the chemical structure of Atoxyl and eventually identified one compound which was the most effective against Syphilis while being the least harmful to humans, which became known as Salvarsan. Salvarsan was widely used to treat Syphilis within years of its discovery.

Fig. 5-6　Paul Ehrlich

(14 March 1854—20 August 1915, aged 61)

Won Nobel Prize in Physiology or Medicine (1908).

5.1.2　Beginning of mass production

In 1928, Alexander Fleming discovered a mold named Penicillium chrysogenum which prevented many types of bacteria from growing. Scientists identified the potential of this mold to provide treatment in humans against bacteria which cause infections. During World War II, the United Kingdom and the United States worked together to find a method of mass producing Penicillin, a derivative of the Penicillium mold, which had the potential to save many lives during the war since it could treat infections common in injured soldiers. Although Penicillin could be isolated from the mold in a laboratory setting, there was no known way to obtain the amount of medication needed to treat the quantity of people who needed it. Scientists with major chemical companies such as Pfizer were able to develop a deep-fermentation process which could produce a high yield of penicillin. In 1944, Pfizer opened the first penicillin factory, and its products were exported to aid the war efforts overseas.

Fig. 5-7　Sir Alexander Fleming

(6 August 1881—11 March 1955, aged 73) Known for Discovery of Penicillin and Lysozyme, Nobel Prize (1945).

5.1.3　Controlled drug release

Tablets for oral consumption of medication have been utilized since approximately 1500 B.C., however for a long time the only method of drug release was immediate dissolution, meaning all of the medication is released in the body at once. In the 1950s, sustained release technology was developed. Through mechanisms such as osmosis and diffusion, pills were designed that could release the medication over a 12-hour to 24-hour period. Smith, Kline & French developed one of the first major successful sustained release technologies. Their formulation consisted of a collection of small tablets taken at the same time, with varying amounts of wax coating that allowed some tablets to dissolve in the body faster than others. The result was a continuous release of the drug as it travelled through the intestinal tract. Although modern day research focuses on extending the controlled release timescale to the order of months, once-a-day and twice-a-day pills are still the most widely utilized controlled drug release method.

Fig. 5-8 Equipment for deep-fermentation of penicillin.

5.1.4 Formation of the ISPE

In 1980, the International Society for Pharmaceutical Engineering (ISPE) was formed to support and guide professionals in the pharmaceutical industry through all parts of the process of bringing new medications to the market. The ISPE writes standards and guidelines for individuals and companies to use and to model their practices after. The ISPE also hosts training sessions and conferences for professionals to attend, learn, and collaborate with others in the field.

5.2 Pharmaceutical industry in the United Kingdom

The pharmaceutical industry in the United Kingdom directly employs around 73,000 people and in 2007 contributed £8.4 billion to the UK's GDP and invested a total of £3.9 billion in research and development. In 2007 exports of pharmaceutical products from the UK totalled £14.6 billion, creating a trade surplus in pharmaceutical products of £4.3 billion.

UK Pharmaceutical employment of 73,000 in 2017 compares to 114,000 as of 2015 in Germany, 92,000 as of 2014 in France and 723,000 in the European Union as a whole. In the United States 281, 440 people work in pharmaceutical industry as of 2016.

The UK is home to GlaxoSmithKline and AstraZeneca, respectively the world's fifth- and sixth-largest pharmaceutical companies measured by 2009 market share. Foreign companies with a major presence in the UK pharmaceutical industry include Pfizer, Novartis, Hoffmann–La Roche and Eisai. One in five of the world's biggest-selling prescription drugs were developed in the UK.

5.2.1 History

1. 19th century

In 1842 Thomas Beecham established the Beecham's Pills laxative business, which would later become the Beecham Group. By 1851 UK-based patent medicine companies had combined domestic revenues of around £250,000. Beecham opened Britain's first modern drugs factory in St Helens in 1859. Henry Wellcomeand Silas Burroughs formed a partnership in September 1880, and established an office in Snow Hill in Central London. The London Wholesale Drug and Chemical Protection Society was formed in 1867, which became the Drug Club in 1891, the forerunner of the present-day Association of the British Pharmaceutical Industry. In 1883 Burroughs Wellcome & Co. opened their first factory, at Bell Lane Wharf in Wandsworth, utilising compressed medicine tablet-making machinery acquired from Wyeth of the United States. Burroughs Wellcome & Co. established its first overseas branch in Sydney in 1898.

Fig. 5-9 Beecham's Clock Tower, constructed in 1877 as part of the Beecham factory in St Helens.

2. 20th century

The Glaxo department of Joseph Nathan and Co. was established in London in 1908. Glaxo Laboratories Ltd. absorbed Joseph Nathan and Co. in 1947 and was listed on the London Stock Exchange in the same year. In order to satisfy regulations then in place in the UK on the importation of medicines, Pfizer established a compounding operation in Folkestone, Kent in Autumn 1952. Pfizer acquired an 80-acre site on the outskirts of Sandwich in 1954 to enable the expansion of its Kent-based activities. Glaxo acquired Allen and Hanburys Ltd. in 1958. In 1981 the bacterial infection treatment Augmentin (amoxicillin/clavulanate potassium) was launched by Beecham; the anti-ulcer treatment Zantac (ranitidine) was launched by Glaxo; and the antiviral herpes treatment Zovirax (aciclovir) was launched by Wellcome.

In 1991 SmithKline Beecham launched Seroxat/Paxil (paroxetine hydrochloride). In June 1993 Imperial Chemical Industries demerged its pharmaceuticals and agrochemicals businesses, forming Zeneca Group plc. In 1995 Glaxo opened a major research and development facility in Stevenage, constructed at a cost of £700 million. In March 1995 the £9 billion acquisition of Wellcome by Glaxo was completed, forming Glaxo Wellcome, in what was the largest merger in UK corporate history to date. BASF completed the acquisition of the pharmaceutical division of The Boots Company in April 1995. In 1997 SmithKline Beecham opened a major new research centre at New Frontiers Science Park in Harlow, Essex. In 1999 Zeneca Group plc and Sweden-based Astra AB merged to form AstraZeneca plc. Glaxo Wellcome and SmithKline Beecham announced their intention to merge in January 2000, with the merger completing in December of that year, forming GlaxoSmithKline plc.

3. 21st century

In February 2001 the Novartis Respiratory Research Centre, the largest single-site respiratory research centre in the world, opened in Horsham. In May 2006 AstraZeneca agreed to buy Cambridge Antibody Technology, then the largest UK-based biotechnology company, for £702 million. In April 2007 AstraZeneca agreed to acquire the U.S.-based biotechnology company MedImmune for $15.6 billion. In April 2009 GlaxoSmithKline agreed to acquire Stiefel Laboratories, then the world's largest independent dermatology company, for US$3.6 billion. In June 2009 Eisai opened a major new research and development and manufacturing facility in Hatfield, constructed at a cost of over £100 million. In November 2009 GlaxoSmithKline and Pfizer combined their respective AIDS divisions into one London-based company, ViiV Healthcare. On 1 February 2011 Pfizer announced that it would be closing its entire research and development facility at Sandwich, Kent within 18~24 months with the loss of 2,400 jobs, as part of a company-wide plan to reduce its spending on research and development.

Fig. 5-10 The world headquarters of GlaxoSmithKline in Brentford, London.

In March 2013 AstraZeneca announced plans for a major corporate restructuring, including the closure of its research and development activities at Alderley Park, investment of $500 million in the construction of a new research and development facility in Cambridge, and the move of its corporate headquarters from London to Cambridge in 2016.

The amount of funding received by UK life science companies reached a 10-year high in 2014.

5.2.2　Research and development

In 2007 the UK had the third-highest share of global pharmaceutical Research and Development (R&D) expenditure of any nation, with 9% of the total, behind the United States (49%) and Japan (15%). The UK has the largest pharmaceutical R&D expenditure of any European nation, accounting for 23% of the total; followed by France (20%), Germany (19%), and Switzerland (11%).

New words

regulatory ['regjələtərɪ] *adj.* 调整的；具有监管权的，监管的
biotechnology [,baɪəʊtek'nɒlədʒɪ] *n.* [生物]生物工艺学；[生物]生物技术
biotechnologist [baɪətek'nɒlədʒɪst] *n.* 生物工艺学家
uphold [ʌp'həʊld] *v.* 支持；维持；赞成；支撑
validate ['vælɪdeɪt] *vt.* 使合法化，使有法律效力；使生效；批准，确认；证实
integrity [ɪn'tegrətɪ] *n.* 正直，诚实；完整；[计算机]保存；健全
facility [fə'sɪlətɪ] *n.* 设备；容易；能力；灵巧
utility [juː'tɪlətɪ] *n.* 功用，效用；有用的物体或器械；公用事业公司；公用事业
　　　　　　adj. 有多种用途的；各种工作都会做的；能在数个位置作替补的；为获得经济价值而饲养
bioprocess ['baɪɒprəʊses] *n.* 生物过程
bioprocessing ['baɪɒprəʊsesɪŋ] *n.* [医]生物工艺
derivative [dɪ'rɪvətɪv] *n.* [数]导数，微商；[化] 衍生物，派生物；[语]派生词
　　　　　　adj. 衍生的；导出的；拷贝的
manipulate [mə'nɪpjʊleɪt] *vt.* 操作，处理；巧妙地控制；操纵；[医] 推拿，调整
arsenic ['ɑːsnɪk] *n.* 砷；三氧化二砷，砒霜
　　　　　　adj. 砷的，含砷（主要指五价砷）的
Treponema [,trepə'nɪːmə] *n.* 密螺旋体
pallidum ['pælɪːdəm] *n.* 苍白球
Treponema pallidum [医]苍白密螺旋体
Syphilis ['sɪfɪlɪs] *n.* 梅毒
parasitic [,pærə'sɪtɪk] *adj.* 寄生的；寄生物的；由寄生虫引起的

adverse ['ædvɜːs] *adj.* 不利的；有害的；逆的；相反的
Salvarsan ['sælvəsən] *n.* 六〇六，洒尔佛散
mold [məʊld] *n.* 模子；模式；类型；霉
　　　　　 vt. 塑造；浇铸；用模子做；用泥土覆盖
　　　　　 vi. 对……产生影响，形成；发霉
chrysogenum [krɪsə'dʒenʌm] 产黄青霉素菌
chrysogenin [krɪsə'dʒenɪn] *n.* 黄青霉素，点青霉产黄色素
fermentation [ˌfɜːmen'teɪʃn] *n.* 发酵；激动，纷扰
ferment [fə'ment] *n.* 酶，酵素；发酵剂；骚动，动乱
　　　　　 vt. & vi. 使发酵；使骚动；酝酿
sustain [sə'steɪn] *vt.* 维持；支撑，支持；遭受，忍受；供养
osmosis [ɒz'məʊsɪs] *n.* 渗透（作用），潜移默化，耳濡目染
intestinal [ˌɪntes'taɪnl] *adj.* 肠的；肠壁；肠道细菌
intestine [ɪn'testɪn] *n.* [解]肠
　　　　　 adj. 内部的；国内的
collaborate [kə'læbəreɪt] *vi.* 合作，协作；（国家间的）协调，提携；勾结，通敌
laxative ['læksətɪv] *n.* 泻药；[药]缓泻药
　　　　　 adj. 通便的
patent ['pætnt] *n.* 专利；专利品；专利权；专利证
　　　　　 adj. 专利的；显然，显露；明摆着的
　　　　　 vt. 获得……专利，给予……专利权；取得专利权
revenue ['revənjuː] *n.* （国家的）岁入，税收，（土地，财产等的）收入，收益，所得，
　　　　　 （个人的）固定收入；[复数]总收入；税务署
forerunner ['fɔːrʌnə(r)] *n.* 先驱；先驱者；先兆，预兆
utilise ['juːtɪlaɪz] *vt.* 利用，使用
augmentin [ɔːg'mentɪn] *n.* <医> 沃格孟汀
augment [ɔːg'ment] *vt.* 增强，加强；增加，增添；（使）扩张，扩大
　　　　　 n. 增加，补充物
ulcer ['ʌlsə(r)] *n.* <医> 溃疡；腐烂物；道德败坏
ranitidine [rə'nɪtɪdiːn] *n.* <医> 雷尼替丁，甲胺呋硫
antiviral [ˌæntɪ'vaɪrəl] *adj.* 抗病毒的
herpes ['hɜːpiːs] <医> 疱疹；疹病毒
aciclovir [æ'sɪkləʊvə] *n.* 阿昔洛韦
demerge [ˌdɪ'mɜːdʒ] *vt.* 使分离，使拆分
agrochemicals [æg'rɒkemɪklz] *n.* <化> 农用化学品
merger ['mɜːdʒə(r)] *n.* （两个公司的）合并；联合体；吸收
respiratory [rə'spɪrətrɪ] *adj.* 呼吸的
dermatology [ˌdɜːmə'tɒlədʒɪ] *n.* 皮肤病（学）；皮肤科
restructure [ˌriː'strʌktʃə] *v.* 重建；调整；重组

Unit 6　Pharmacy

Pharmacy is the science and technique of preparing, dispensing, and review of drugs and providing additional clinical services. It is a health profession that links health sciences with pharmaceutical sciences and aims to ensure the safe, effective, and affordable use of drugs. The professional practice is becoming more clinically oriented as most of the drugs are now manufactured by pharmaceutical industries. Based on the setting, the pharmacy is classified as a community or institutional pharmacy. Providing direct patient care in the community of institutional pharmacies are considered clinical pharmacy.

The scope of pharmacy practice includes more traditional roles such as compounding and dispensing medications, and it also includes more modern services related to health care, including clinical services, reviewing medications for safety and efficacy, and providing drug information. Pharmacists, therefore, are the experts on drug therapy and are the primary health professionals who optimize use of medication for the benefit of the patients.

Fig. 6-1　Pharmacy, tacuinum sanitatis casa natensis (14th century).

An establishment in which pharmacy (in the first sense) is practiced is called a pharmacy (this term is more common in the United States) or a chemist's (which is more common in Great Britain). In the United States and Canada, drugstores commonly sell medicines, as well as miscellaneous items such as confectionery, cosmetics, office supplies, toys, hair care products and magazines and occasionally refreshments and groceries.

In its investigation of herbal and chemical ingredients, the work of the pharma may be regarded as a precursor of the modern sciences of chemistry and pharmacology, prior to the formulation of the scientific method.

6.1 Disciplines

The field of pharmacy can generally be divided into three primary disciplines:
(1) Pharmaceutics.
(2) Medicinal chemistry and pharmacognosy.
(3) Pharmacy practice.

The boundaries between these disciplines and with other sciences, such as biochemistry, are not always clear-cut. Often, collaborative teams from various disciplines (pharmacists and other scientists) work together toward the introduction of new therapeutics and methods for patient care. However, pharmacy is not a basic or biomedical science in its typical form. Medicinal chemistry is also a distinct branch of synthetic chemistry combining pharmacology, organic chemistry, and chemical biology.

Pharmacology is sometimes considered as the 4th discipline of pharmacy. Although pharmacology is essential to the study of pharmacy, it is not specific to pharmacy. Both disciplines are distinct. Those who wish to practice both pharmacy (patient oriented) and pharmacology (a biomedical science requiring the scientific method) receive separate training and degrees unique to either discipline.

Pharmacoinformatics is considered another new discipline, for systematic drug discovery and development with efficiency and safety.

6.2 Pharmacy technicians

Pharmacy technicians support the work of pharmacists and other health professionals by performing a variety of pharmacy related functions, including dispensing prescription drugs and other medical devices to patients and instructing on their use. They may also perform administrative duties in pharmaceutical practice, such as reviewing prescription requests with medic's offices and insurance companies to ensure correct medications are provided and payment is received.

A pharmacy technician in the UK has recently been referred to by some as a professional. Legislation requires the supervision of certain pharmacy technician's activities by a pharmacist. The majority of pharmacy technicians work in community pharmacies. In hospital pharmacies, pharmacy technicians may be managed by other senior pharmacy technicians. In the UK the role of a PhT in hospital pharmacy has grown and responsibility has been passed on to them to manage the pharmacy department and specialised areas in pharmacy practice allowing pharmacists the time to specialise in their expert field as medication consultants spending more time working with patients and in research. Pharmacy technicians are registered with the General Pharmaceutical Council (GPhC). The GPhC is the regulator of pharmacists, pharmacy technicians and pharmacy premises.

In the US, pharmacy technicians perform their duties under supervision of pharmacists. Although they may perform, under supervision, most dispensing, compounding and other tasks, they are not generally allowed to perform the role of counseling patients on the proper use of their medications.

6.3 Education requirements

There are different requirements of schooling based on the area of pharmaceuticals a student is seeking. In the United States, the general pharmacist will attain a Doctor of Pharmacy Degree (Pharm. D.). The Pharm. D. can be completed in a minimum of six years, which includes two years of pre-pharmacy classes, and four years of professional studies. After graduating pharmacy school, it is highly suggested that the student go on to complete a one or two year residency, which provides valuable experience for the student before going out independently to be a generalized or specialized pharmacist.

The curriculum created for a Pharm. D. is made up of 208-credit hours. Of the 208-credit hours, 68 are transferred-credit hours, and the remaining 140-credit hours are completed in the professional school. There are a series of required standardized tests that students have to pass throughout the process of pharmacy school. The standardized test to get into pharmacy school is called the Pharmacy College Admission Test (PCAT). In a student's third professional year in pharmacy school, it is required to pass the Pharmacy Curriculum Outcomes Assessment (PCOA). Once the Pharm. D. is attained after the fourth year professional school, the student is then eligible to take the North American Pharmacist Licensure Exam (NAPLEX) and the Multistate Pharmacy Jurisprudence Exam (MPJE) to work as a professional pharmacist.

6.4 History

The earliest known compilation of medicinal substances was the Sushruta Samhita, an Indian Ayurvedic treatise attributed to Sushruta in the 6th century BC. However, the earliest text as preserved dates to the 3rd or 4th century AD.

Many Sumerian (late 6th millennium BC — early 2nd millennium BC) cuneiform clay tablets record prescriptions for medicine.

Fig. 6-2 Physician and Pharmacist, illustration from Medicinarius (1505) by Hieronymus Brunschwig.

Ancient Egyptian pharmacological knowledge was recorded in various papyri such as the Ebers Papyrus of 1550 BC, and the Edwin Smith Papyrus of the 16th century BC.

In Ancient Greece, Diocles of Carystus (4th century BC) was one of several men studying the medicinal properties of plants. He wrote several treatises on the topic. The Greek physician Pedanius Dioscorides is famous for writing a five volume book in his native Greek Περί ύλης ιατρικής in the 1st century AD. The Latin translation De Materia Medica (Concerning medical substances) was used a basis for many medieval texts, and was built upon by many middle eastern scientists during the Islamic Golden Age.

Fig. 6-3 Dioscorides, De Materia Medica, Byzantium, 15th century.

Pharmacy in China dates at least to the earliest known Chinese manual, the *Shennong Bencao Jing* (*The Divine Farmer's Herb-Root Classic*), dating back to the 1st century AD. It was compiled during the Han dynasty and was attributed to the mythical Shennong. Earlier literature included lists of prescriptions for specific ailments, exemplified by a manuscript "Recipes for 52 Ailments", found in the Mawangdui, sealed in 168 BC.

In Japan, at the end of the Asuka period (538–710) and the early Nara period (710–794), the men who fulfilled roles similar to those of modern pharmacists were highly respected. The place of pharmacists in society was expressly defined in the Taihō Code (701) and re-stated in the Yōrō Code (718). Ranked positions in the pre-Heian Imperial court were established; and this organizational structure remained largely intact until the Meiji Restoration (1868). In this highly stable hierarchy, the pharmacists—and even pharmacist assistants—were assigned status superior to all others in health-related fields such as physicians and acupuncturists. In the Imperial household, the pharmacist was even ranked above the two personal physicians of the Emperor.

There is a stone sign for a pharmacy with a tripod, a mortar, and a pestle opposite one for a doctor in the Arcadian Way in Ephesus near Kusadasi in Turkey. The current Ephesus dates back to 400 BC and was the site of the Temple of Artemis, one of the seven wonders of the world.

In Baghdad the first pharmacies, or drug stores, were established in 754, under the Abbasid Caliphate during the Islamic Golden Age. By the 9th century, these pharmacies were state-regulated.

The advances made in the Middle East in botany and chemistry led medicine in medieval Islam substantially to develop pharmacology. Muhammad ibn Zakarīya Rāzi(Rhazes) (865–915), for instance, acted to promote the medical uses of chemical compounds. Abu al-Qasim al-Zahrawi

(Abulcasis) (936–1013) pioneered the preparation of medicines by sublimation and distillation. His Liber servitoris is of particular interest, as it provides the reader with recipes and explains how to prepare the "simples" from which were compounded the complex drugs then generally used. Sabur Ibn Sahl (d. 869), was, however, the first physician to initiate pharmacopoeia, describing a large variety of drugs and remedies for ailments. Al-Biruni (973–1050) wrote one of the most valuable Islamic works on pharmacology, entitled Kitab al-Saydalah (*The Book of Drugs*), in which he detailed the properties of drugs and outlined the role of pharmacy and the functions and duties of the pharmacist. Avicenna, too, described no less than 700 preparations, their properties, modes of action, and their indications. He devoted in fact a whole volume to simple drugs in *The Canon of Medicine*. Of great impact were also the works by al-Maridini of Baghdad and Cairo, and Ibn al-Wafid (1008–1074), both of which were printed in Latin more than fifty times, appearing as De Medicinis universalibus et particularibus by "Mesue" the younger, and the Medicamentis simplicibus by "Abenguefit". Peter of Abano (1250–1316) translated and added a supplement to the work of al-Maridini under the title De Veneris. Al-Muwaffaq's contributions in the field are also pioneering. Living in the 10th century, he wrote *The Foundations of the True Properties of Remedies*, amongst others describing arsenious oxide, and being acquainted with silicic acid. He made clear distinction between sodium carbonate and potassium carbonate, and drew attention to the poisonous nature of copper compounds, especially copper vitriol, and also lead compounds. He also describes the distillation of sea-water for drinking.

In Europe pharmacy-like shops began to appear during the 12th century. In 1240 emperor Frederic II issued a decree by which the physician's and the apothecary's professions were separated. "The first pharmacy in Europe (still working) was opened in 1241 in Trier, Germany."

In Europe there are old pharmacies still operating in Dubrovnik, Croatia, located inside the Franciscan monastery, opened in 1317; and in the Town Hall Square of Tallinn, Estonia, dating from at least 1422. The oldest is claimed to have been set up in 1221 in the Church of Santa Maria Novella in Florence, Italy, which now houses a perfume museum. The medieval Esteve Pharmacy, located in Llívia, a Catalan enclave close to Puigcerdà, also now a museum, dates back to the 15th century, keeping albarellos from the 16th and 17th centuries, old prescription books and antique drugs.

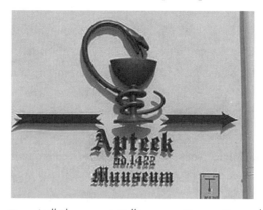

Fig. 6-4 Sign of the Town Hall Pharmacy in Tallinn, operating continuously from at least 1422.

6.5 Practice areas

Pharmacists practice in a variety of areas including community pharmacies, hospitals, clinics, extended care facilities, psychiatric hospitals, and regulatory agencies. Pharmacists themselves may have expertise in a medical specialty.

6.5.1 Community pharmacy

A pharmacy (commonly the chemist in Australia, New Zealand and the UK; or drugstore in North America; retail pharmacy in industry terminology; or Apothecary, historically) is the place where most pharmacists practice the profession of pharmacy. It is the community pharmacy where the dichotomy of the profession exists — health professionals who are also retailers.

Community pharmacies usually consist of a retail storefront with a dispensary where medications are stored and dispensed. According to Sharif Kaf al-Ghazal, the opening of the first drugstores are recorded by Muslim pharmacists in Baghdad in 754.

In most countries, the dispensary is subject to pharmacy legislation; with requirements for storage conditions, compulsory texts, equipment, etc., specified in legislation. Where it was once the case that pharmacists stayed within the dispensary compounding/dispensing medications, there has been an increasing trend towards the use of trained pharmacy technicians while the pharmacist spends more time communicating with patients. Pharmacy technicians are now more dependent upon automation to assist them in their new role dealing with patients' prescriptions and patient safety issues.

Fig. 6-5 19th-century Italian pharmacy.

Pharmacies are typically required to have a pharmacist on-duty at all times when open. It is also often a requirement that the owner of a pharmacy must be a registered pharmacist, although this is not the case in all jurisdictions, such that many retailers (including supermarkets and mass merchandisers) now include a pharmacy as a department of their store.

Likewise, many pharmacies are now rather grocery store-like in their design. In addition to medicines and prescriptions, many now sell a diverse arrangement of additional items such as cosmetics, shampoo, office supplies, confections, snack foods, durable medical equipment, greeting cards, and provide photo processing services.

Fig. 6-6 Classic symbols at the wall of a former German pharmacy.

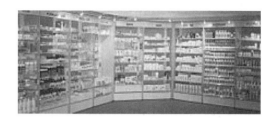

Fig. 6-7 Modern pharmacy in Norway.

6.5.2 Hospital pharmacy

Pharmacies within hospitals differ considerably from community pharmacies. Some pharmacists in hospital pharmacies may have more complex clinical medication management issues whereas pharmacists in community pharmacies often have more complex business and customer relations issues.

Because of the complexity of medications including specific indications, effectiveness of treatment regimens, safety of medications (i.e., drug interactions) and patient compliance issues (in the hospital and at home) many pharmacists practicing in hospitals gain more education and training after pharmacy school through a pharmacy practice residency and sometimes followed by another residency in a specific area. Those pharmacists are often referred to as clinical pharmacists and they often specialize in various disciplines of pharmacy. For example, there are pharmacists who specialize in hematology/oncology, HIV/AIDS, infectious disease, critical care, emergency medicine, toxicology, nuclear pharmacy, pain management, psychiatry, anti-coagulation clinics, herbal medicine, neurology/epilepsy management, pediatrics, neonatal pharmacists and more.

Hospital pharmacies can often be found within the premises of the hospital. Hospital pharmacies usually stock a larger range of medications, including more specialized medications, than would be feasible in the community setting. Most hospital medications are unit-dose, or a single dose of medicine. Hospital pharmacists and trained pharmacy technicians compound sterile products for patients including total parenteral nutrition (TPN), and other medications given intravenously. This is a complex process that requires adequate training of personnel, quality assurance of products, and adequate facilities. Several hospital pharmacies have decided to outsource high risk preparations and some other compounding functions to companies who specialize in compounding. The high cost of medications and drug-related technology, combined with the potential impact of medications and pharmacy services on patient-care outcomes and patient safety, make it imperative that hospital pharmacies perform at the highest level possible.

6.5.3 Clinical pharmacy

Pharmacists provide direct patient care services that optimizes the use of medication and promotes health, wellness, and disease prevention. Clinical pharmacists care for patients in all health care settings, but the clinical pharmacy movement initially began inside hospitals and clinics. Clinical pharmacists often collaborate with physicians and other healthcare professionals to improve pharmaceutical care. Clinical pharmacists are now an integral part of the interdisciplinary approach to patient care. They often participate in patient care rounds for drug product selection.

The clinical pharmacist's role involves creating a comprehensive drug therapy plan for patient-specific problems, identifying goals of therapy, and reviewing all prescribed medications prior to dispensing and administration to the patient. The review process often involves an evaluation of the appropriateness of the drug therapy (e.g., drug choice, dose, route, frequency, and duration of therapy) and its efficacy. The pharmacist must also monitor for potential drug interactions, adverse drug reactions, and assess patient drug allergies while designing and initiating a drug therapy plan.

6.5.4 Ambulatory care pharmacy

Since the emergence of modern clinical pharmacy, ambulatory care pharmacy practice has emerged as a unique pharmacy practice setting. Ambulatory care pharmacy is based primarily on pharmacotherapy services that a pharmacist provides in a clinic. Pharmacists in this setting often do not dispense drugs, but rather see patients in office visits to manage chronic disease states.

In the U.S. federal health care system (including the VA, the Indian Health Service, and NIH) ambulatory care pharmacists are given full independent prescribing authority. In some states such North Carolina and New Mexico these pharmacist clinicians are given collaborative prescriptive and diagnostic authority. In 2011 the board of Pharmaceutical Specialties approved ambulatory care pharmacy practice as a separate board certification. The official designation for pharmacists who pass the ambulatory care pharmacy specialty certification exam will be Board Certified Ambulatory Care Pharmacist (BCACP) and these pharmacists will carry the initials BCACP.

6.5.5 Compounding pharmacy

Compounding is the practice of preparing drugs in new forms. For example, if a drug manufacturer only provides a drug as a tablet, a compounding pharmacist might make a medicated lollipop that contains the drug. Patients who have difficulty swallowing the tablet may prefer to suck the medicated lollipop instead.

Another form of compounding is by mixing different strengths (g, mg, mcg) of capsules or tablets to yield the desired amount of medication indicated by the physician, physician assistant, Nurse Practitioner, or clinical pharmacist practitioner. This form of compounding is found at community or hospital pharmacies or in-home administration therapy.

Compounding pharmacies specialize in compounding, although many also dispense the same non-compounded drugs that patients can obtain from community pharmacies.

6.5.6　Consultant pharmacy

Consultant pharmacy practice focuses more on medication regimen review (i.e. "cognitive services") than on actual dispensing of drugs. Consultant pharmacists most typically work in nursing homes, but are increasingly branching into other institutions and non-institutional settings. Traditionally consultant pharmacists were usually independent business owners, though in the United States many now work for several large pharmacy management companies (primarily Omnicare, Kindred Healthcare and PharMerica). This trend may be gradually reversing as consultant pharmacists begin to work directly with patients, primarily because many elderly people are now taking numerous medications but continue to live outside of institutional settings. Some community pharmacies employ consultant pharmacists and/or provide consulting services.

The main principle of consultant pharmacy is developed by Hepler and Strand in 1990.

6.5.7　Internet pharmacy

Since about the year 2000, a growing number of Internet pharmacies have been established worldwide. Many of these pharmacies are similar to community pharmacies, and in fact, many of them are actually operated by brick-and-mortar community pharmacies that serve consumers online and those that walk in their door. The primary difference is the method by which the medications are requested and received. Some customers consider this to be more convenient and private method rather than traveling to a community drugstore where another customer might overhear about the drugs that they take. Internet pharmacies (also known as online pharmacies) are also recommended to some patients by their physicians if they are homebound.

Fig. 6-8　Canisters of pills from a mail order pharmacy.

While most Internet pharmacies sell prescription drugs and require a valid prescription, some Internet pharmacies sell prescription drugs without requiring a prescription. Many customers order drugs from such pharmacies to avoid the "inconvenience" of visiting a doctor or to obtain medications which their doctors were unwilling to prescribe. However, this practice has been criticized as potentially dangerous, especially by those who feel that only doctors can reliably assess contraindications, risk/benefit ratios, and an individual's overall suitability for use of medication. There also have been reports of such pharmacies dispensing substandard products.

Of particular concern with Internet pharmacies is the ease with which people, youth in particular, can obtain controlled substances (e.g., Vicodin, generically known as hydrocodone) via the Internet without a prescription issued by a doctor/practitioner who has an established doctor-patient relationship. There are many instances where a practitioner issues a prescription, brokered by an Internet server, for a controlled substance to a "patient" s/he has never met. In the United States, in order for a prescription for a controlled substance to be valid, it must be issued for a legitimate medical purpose by a licensed practitioner acting in the course of legitimate doctor-patient relationship. The filling pharmacy has a corresponding responsibility to ensure that the prescription is valid. Often, individual state laws outline what defines a valid patient-doctor relationship.

Canada is home to dozens of licensed Internet pharmacies, many of which sell their lower-cost prescription drugs to U.S. consumers, who pay one of the world's highest drug prices. In recent years, many consumers in the US and in other countries with high drug costs, have turned to licensed Internet pharmacies in India, Israel, and the UK, which often have even lower prices than in Canada.

In the United States, there has been a push to legalize the importation of medications from Canada and other countries, in order to reduce consumer costs. While in most cases importation of prescription medications violates Food and Drug Administration (FDA) regulations and federal laws, enforcement is generally targeted at international drug suppliers, rather than consumers. There is no known case of any U.S. citizens buying Canadian drugs for personal use with a prescription, who has ever been charged by authorities.

6.5.8 Veterinary pharmacy

Veterinary pharmacies, sometimes called animal pharmacies, may fall in the category of hospital pharmacy, retail pharmacy or mail-order pharmacy. Veterinary pharmacies stock different varieties and different strengths of medications to fulfill the pharmaceutical needs of animals. Because the needs of animals, as well as the regulations on veterinary medicine, are often very different from those related to people, veterinary pharmacy is often kept separate from regular pharmacies.

6.5.9　Nuclear pharmacy

Nuclear pharmacy focuses on preparing radioactive materials for diagnostic tests and for treating certain diseases. Nuclear pharmacists undergo additional training specific to handling radioactive materials, and unlike in community and hospital pharmacies, nuclear pharmacists typically do not interact directly with patients.

6.5.10　Military pharmacy

Military pharmacy is an entirely different working environment due to the fact that technicians perform most duties that in a civilian sector would be illegal. State laws of technician patient counseling and medication checking by a pharmacist do not apply.

6.5.11　Pharmacy informatics

Pharmacy informatics is the combination of pharmacy practice science and applied information science. Pharmacy informaticists work in many practice areas of pharmacy, however, they may also work in information technology departments or for healthcare information technology vendor companies. As a practice area and specialist domain, pharmacy informatics is growing quickly to meet the needs of major national and international patient information projects and health system interoperability goals. Pharmacists in this area are trained to participate in medication management system development, deployment and optimization.

6.5.12　Specialty pharmacy

Specialty pharmacies supply high cost injectable, oral, infused, or inhaled medications that are used for chronic and complex disease states such as cancer, hepatitis, and rheumatoid arthritis. Unlike a traditional community pharmacy where prescriptions for any common medication can be brought in and filled, specialty pharmacies carry novel medications that need to be properly stored, administered, carefully monitored, and clinically managed. In addition to supplying these drugs, specialty pharmacies also provide lab monitoring, adherence counseling, and assist patients with cost-containment strategies needed to obtain their expensive specialty drugs. It is currently the fastest growing sector of the pharmaceutical industry with 19 of 28 newly FDA approved medications in 2013 being specialty drugs.

Due to the demand for clinicians who can properly manage these specific patient populations, the Specialty Pharmacy Certification Board has developed a new certification exam to certify specialty pharmacists. Along with the 100 question computerized multiple-choice exam, pharmacists must also complete 3,000 hours of specialty pharmacy practice within the past three years as well as 30 hours of specialty pharmacist continuing education within the past two years.

6.6 The future of pharmacy

In the coming decades, pharmacists are expected to become more integral within the health care system. Rather than simply dispensing medication, pharmacists are increasingly expected to be compensated for their patient care skills. In particular, Medication Therapy Management (MTM) includes the clinical services that pharmacists can provide for their patients. Such services include the thorough analysis of all medication (prescription, non-prescription, and herbals) currently being taken by an individual. The result is a reconciliation of medication and patient education resulting in increased patient health outcomes and decreased costs to the health care system.

This shift has already commenced in some countries; for instance, pharmacists in Australia receive remuneration from the Australian Government for conducting comprehensive Home Medicines Reviews. In Canada, pharmacists in certain provinces have limited prescribing rights (as in Alberta and British Columbia) or are remunerated by their provincial government for expanded services such as medications reviews (Medschecks in Ontario). In the United Kingdom, pharmacists who undertake additional training are obtaining prescribing rights and this is because of pharmacy education. They are also being paid for by the government for medicine use reviews. In Scotland the pharmacist can write prescriptions for Scottish registered patients of their regular medications, for the majority of drugs, except for controlled drugs, when the patient is unable to see their doctor, as could happen if they are away from home or the doctor is unavailable. In the United States, pharmaceutical care or clinical pharmacy has had an evolving influence on the practice of pharmacy. Moreover, the Doctor of Pharmacy (Pharm. D.) degree is now required before entering practice and some pharmacists now complete one or two years of residency or fellowship training following graduation. In addition, consultant pharmacists, who traditionally operated primarily in nursing homes are now expanding into direct consultation with patients, under the banner of "senior care pharmacy".

In addition to patient care, pharmacies will be a focal point for medical adherence initiatives. There is enough evidence to show that integrated pharmacy based initiatives significantly impact adherence for chronic patients. For example, a study published in NIH shows "pharmacy based interventions improved patients' medication adherence rates by 2.1 percent and increased physicians' initiation rates by 38 percent, compared to the control group".

6.7 Symbols

The two symbols most commonly associated with pharmacy in English-speaking countries are the mortar and pestle and the ℞ (recipere) character, which is often written as "Rx" in typed text. The show globe was also used until the early 20th century. Pharmacy organizations often use other symbols, such as the Bowl of Hygieia which is often used in the Netherlands, conical measures, and caduceusesin their logos. Other symbols are common in different countries: the green Greek cross in France, Argentina, the United Kingdom, Belgium, Ireland, Italy, Spain, and India, the increasingly rare Gaper in the Netherlands, and a red stylized letter "A" in Germany and Austria (from Apotheke, the German word for pharmacy, from the same Greek root as the English word "apothecary").

Fig. 6-9　The mortar and pestle, used in the United States and Canada.

Fig. 6-10　The symbol used on medical prescriptions, from the Latin Recipe.

Fig. 6-11　19th Century Apothecary Globe Pendant.

Fig. 6-12　Green cross and Bowl of Hygieia used in Europe (with the exception of Germany and Austria) and India.

Fig. 6-13　Simple green cross, also used in Europe and India.

Fig. 6-14　Gaper on the front of "Van der Pigge", a pharmacy in Haarlem.

Fig. 6-15　Red "A" (Apotheke) sign, used in Germany.

Fig. 6-16　Similar red "A" sign, used in Austria.

Fig. 6-17　Rod of Asclepius, the internationally recognised symbol of medicine.

New words

pharmacy ['fɑːməsi] *n.* 药房；配药学，药学；制药业；一批备用药品
dispense [dɪ'spens] *vt.* 分配，分给；实施，施行；免除，豁免；配（药）
　　　　　　　　　　vi. 特许，豁免
pharmacist ['fɑːməsɪst] *n.* 药剂师
therapy ['θerəpɪ] *n.* 治疗，疗法，疗效；心理治疗；治疗力
optimize ['ɒptɪmaɪz] *vt.* 使最优化，使尽可能有效
miscellaneous [ˌmɪsə'leɪnɪəs] *adj.* 各种各样的；五花八门的；混杂的；多方面的
confectionery [kən'fekʃənərɪ] *n.* 甜食
refreshment [rɪ'freʃmənt] *n.* 提神，精神恢复；提神物；点心，茶点
herbal ['hɜːbl] *adj.* 药草的；草本的
　　　　　　　　n. 草药书；〈古〉草本志；植物标本
pharma ['fɑːmə] *n.* 医药，制药，医药公司，制药公司
precursor [prɪː'kɜːsə] *n.* 前辈，前驱，先锋，前任；预兆，先兆，初期形式；[核]前驱波，
　　　　　　　　初级粒子；预报器
pharmacology [ˌfɑːmə'kɒlədʒɪ] *n.* 药理学，药物学
pharmaceutics [ˌfɑːmə'sjuːtɪks] *n.* 配药学，制药学；药物学
pharmacognosy [ˌfɑːmə'kɒgnəsɪ] *n.* （尤指研究天然药物的）生药学
theraputics [ˌθerə'pjuːtɪks] *n.* 治疗学，疗法
medicinal [mə'dɪsɪnl] *adj.* 医学的；医药的，药用的；治疗的，医疗的
pharmacoinformatics [ˌfɑːməkɒɪnfə'mætɪks] *n.* 药物信息学
pharmacy technician 药剂师助理；药剂师；药品技师；助理药剂师；药房技术员
medic ['medɪk] *n.* 医生；医学院学生
premise ['premɪs] *n.* 前提；[复数]房屋；[复数][合同、契约用语]上述各点；（逻辑学中的）
　　　　　　　　前提
　　　　　　　　vt. 预述（条件等）；提出……为前提；假设
residency ['rezɪdənsɪ] *n.* 住所；（实习医师的）高级专科住院实习期
eligible ['elɪdʒəbl] *adj.* 合适的；在（法律上或道德上）合格的；有资格当选的；称心如
　　　　　　　　意的
　　　　　　　　n. 合格者；合适者；称心如意的人；合乎条件的人（或东西）
licensure ['laɪsənʃʊə] *n.* 发给许可证
jurisprudence [ˌdʒʊərɪs'pruːdns] *n.* 法学，法理学
compilation [ˌkɒmpɪ'leɪʃn] *n.* 编辑；汇编
treatise ['triːtɪs] *n.* 论文；论述；专著
Sumerian [sʊ'mɪərɪən] *n.* 苏美尔人；苏美尔语
　　　　　　　　adj. 苏美尔的；苏美尔人的
millennium [mɪ'lenɪəm] *n.* 一千年；千年期；千禧年；全人类未来的幸福时代

cuneiform ['kjuːnɪfɔːm] *adj.* 楔形的，楔状骨的，楔形文字的
　　　　　　n. 楔形文字，楔状骨
clay [kleɪ] *n.* 黏土，泥土；（相对于灵魂而言的）人体，肉体；似黏土的东西
　　　　vt. 用黏土处理
tablet ['tæblət] *n.* 碑，匾；药片；便笺簿；小块
　　　　　　vt. 用碑牌纪念；将（备忘录等）写在板上；将……制成小片或小块
papyrus [pə'paɪrəs] *n.*（非洲的）纸莎草；（古代埃及、罗马和希腊用的）纸莎草纸；
　　　　　　（写在纸莎草纸上的）古代文献
papyri [pə'paɪriː]　（papyrus）的复数形式
medieval [ˌmedɪ'iːvl] *adj.* 中古的，中世纪的
manual ['mænjuəl] *adj.* 用手的；手制的，手工的；[法]占有的；体力的
　　　　　　n. 手册；指南；[乐]键盘；[军]刀枪操练
compile [kəm'paɪl] *vt.* 汇编；编辑；编制；编译
mythical ['mɪθɪkl] *adj.* 神话的；虚构的
ailment ['eɪlmənt] *n.* 疾病（尤指微恙）；不安
expressly [ɪk'spreslɪ] *adv.* 明显地；明确地；特别地；特意地
intact [ɪn'tækt] *adj.* 完整无缺的，未经触动的，未受损伤的；原封不动的；完好无缺；
　　　　完好无损
acupuncturist ['ækjupʌŋktʃərɪst] *n.* 针灸师
tripod ['traɪpɒd] *n.* [摄]三脚架；三脚凳；三脚桌；三足鼎
mortar ['mɔːtə(r)] *n.* 迫击炮；砂浆；房产；研钵
pestle ['pesl] *n.* 杵，碾槌
Arcadian [ɑː'keɪdɪən] *adj.* 田园牧歌式的
Artemis ['ɑːtɪmɪs] *n.* 阿耳特弥斯
botany ['bɒtənɪ] *n.* 植物学；精纺毛纱；细羊毛
pharmacopoeia [ˌfɑːməkə'piːə] *n.* 药典，一批备用药品
vitriol ['vɪtrɪɒl] *n.* 硫酸盐，刻薄话；蓝矾
decree [dɪ'kriː] *n.* 法令，命令；（法院的）判决，裁定；（教会的）教令
　　　　vt. 命令；颁布……为法令；（命运）注定；裁决
　　　　vi. 发布命令；注定
Franciscan [fræn'sɪskən] *adj.* 圣方济各会的
　　　　　　n. 方济各会的修士[修女]
monastery ['mɒnəstrɪ] *n.* 修道院，寺院；[复数]全体僧侣
perfume ['pɜːfjuːm] *n.* 香水；香料；香味，香气
　　　　　　vt. 使……充满香气；喷香水于……
enclave ['enkleɪv] *n.* 飞地；被包围的领土
psychiatric [ˌsaɪkɪ'ætrɪk] *adj.* 精神病学的；精神病治疗的
regulatory ['regjələtərɪ] *adj.* 调整的；具有监管权的，监管的
expertise [ˌekspɜː'tiːz] *n.* 专门知识或技能；专家的意见；专家评价，鉴定

retail ['riːteɪl] *n.* 零售
　　　　　　　　vt. 零售；零卖；转述；传播
　　　　　　　　adj. 零售的
　　　　　　　　adv. 以零售方式
apothecary [ə'pɒθəkərɪ] *n.* 药剂师，药店
dichotomy [daɪ'kɒtəmɪ] *n.* 一分成二，对分
storefront ['stɔːfrʌnt] *n.* <美>店面（房），铺面（房），临街房
　　　　　　　　adj. 临街的；在一楼工作的
dispensary [dɪ'spensərɪ] *n.* 医务室，药房
compulsory [kəm'pʌlsərɪ] *adj.* 必须做的，强制性的；义务的；必修的
jurisdiction [ˌdʒʊərɪs'dɪkʃn] *n.* 司法权；管辖权；管辖范围；权限
confection [kən'fekʃn] *n.* 甜食，糕点
photo processing 洗印照片；进行光学处理；照片处理
regimen ['redʒɪmən] *n.* （为病人规定的）生活规则，养生法；养生之道
compliance [kəm'plaɪəns] *n.* 服从，听从；承诺；柔软度；顺度
hematology [ˌhiːmə'tɒlədʒɪ] *n.* 血液学
oncology [ɒŋ'kɒlədʒɪ] *n.* 肿瘤学
toxicology [ˌtɒksɪ'kɒlədʒɪ] *n.* 毒理学，毒物学
coagulation [kəʊˌægjʊ'leɪʃn] *n.* 凝结，凝结物
anti-coagulation 抗凝血
neurology [njʊə'rɒlədʒɪ] *n.* <医>神经病学；<医>神经病学家
　　　　　　　　adj. <医>神经病学的
epilepsy ['epɪlepsɪ] *n.* [医]癫痫，羊癫疯
pediatrics [ˌpiːdɪ'ætrɪks] *n.* 小儿科；儿科学；幼科
neonatal [ˌniːəʊ'neɪtl] *adj.* 新生的，初生的
sterile ['steraɪl] *adj.* 不毛的，贫瘠的；不生育的；无菌的；无效果的
parenteral [pə'rentərəl] *adj.* 肠胃外的；不经肠的；非肠道的；注射用药物的
total parenteral nutrition 胃肠道外全面营养
intravenously [ˌɪntrə'viːnəslɪ] *adv.* 静脉内地，静脉注射地
outsource ['aʊtsɔːs] *vt.* 外购（指从外国供应商等处获得货物或服务）；外包（工程）
imperative [ɪm'perətɪv] *adj.* 必要的，不可避免的；命令的，专横的；势在必行的；
　　　　　　　　[语]祈使的
　　　　　　　　n. 必要的事；命令；规则；[语]祈使语气
integral ['ɪntɪgrəl] *adj.* 完整的；积分的
　　　　　　　　n. 整体；积分
duration [djʊ'reɪʃn] *n.* 持续，持续的时间；（时间的）持续，持久，连续
adverse ['ædvɜːs] *adj.* 不利的；有害的；逆的；相反的
ambulatory ['æmbjələtərɪ] *adj. & n.* 走动的，流动的，非固定的回廊
chronic ['krɒnɪk] *adj.* 慢性的；长期的；习惯性的

lollipop ['lɒlɪpɒp] *n*. 棒棒糖
overhear [,əʊvə'hɪə(r)] *vt*. 偶然听到；无意中听到；偷听
　　　　　　vi. 无意中听到；偷听到
homebound ['həʊmbaʊnd] *adj*. 回家的，回家乡的
contraindication [,kɒntrə,ɪndɪ'keɪʃn] *n*. 禁忌征候
substandard [,sʌb'stændəd] *adj*. 不够标准的，在标准以下的；低等级标准
broker ['brəʊkə] *n*. (股票、外币等)经纪人；中间人，代理人；旧货商人
　　　　　vt. 作为权力经纪人进行谈判；作为中间人来安排、设法
legitimate [lɪ'dʒɪtɪmət] *adj*. 合法的，合理的；正规的；合法婚姻所生的；真正的，真实的
　　　　　　vt. 使合法；给予合法的地位；通过法律手段给（私生子）以合法地位；正式批准，授权
veterinary ['vetrənəri] *adj*. 兽医的
　　　　　　n. 兽医
stock [stɒk] *n*. 股份，股票；库存；树干；家畜
　　　　 adj. 常备的，存货的；陈旧的
　　　　 vt. 提供货物；备有
　　　　 vi. 出新芽；囤积
civilian [sə'vɪlɪən] *n*. 平民，百姓；市民；文官；民法学者
　　　　　 adj. 民用的；平民的
sector ['sektə(r)] *n*. 部门；领域；防御地区；扇形
informatics [,ɪnfə'mætɪks] *n*. 信息学，情报学
vendor ['vendə] *n*. 摊贩，小贩；卖主；[贸易]自动售货机；<正式>供应商
domain [də'meɪn] *n*. 范围，领域；领土，疆土；管辖范围；[计]域名
interoperability ['ɪntərɒpərə'bɪlətɪ] *n*. 互用性，协同工作的能力
injectable [ɪn'dʒektəbl] *n*. 血管注射剂
　　　　　　adj. 可注射的
infuse [ɪn'fjuːz] *vt*. 灌输，加入（一种特性）；使充满；浸渍；鼓舞，激发
　　　　 vi. 沏（茶），泡（草药）
inhale [ɪn'heɪl] *vt. & vi*. 吸入；〈非〉狼吞虎咽；吸气
hepatitis [,hepə'taɪtɪs] *n*. 肝炎
rheumatoid ['ruːmətɔɪd] *adj*. 风湿病的，类风湿病的
arthritis [ɑː'θraɪtɪs] *n*. 关节炎
adherence [əd'hɪərəns] *n*. 依附；坚持；忠诚；密着
containment [kən'teɪnmənt] *n*. 牵制；包含；容量；密闭度
clinician [klɪ'nɪʃn] *n*. 临床医生，门诊医师
reconciliation [,rekənsɪlɪ'eɪʃn] *n*. 和解，调停；一致；服从，顺从；和谐
remuneration [rɪ,mjuːnə'reɪʃn] *n*. 酬报；偿还；酬金；工资
evolve [ɪ'vɒlv] *vt*. 使发展；使进化；设计，制订出；发出，散发
　　　　 vi. 发展；[生]通过进化进程发展或发生

focal ['fəʊkl] *adj.* 焦点的，有焦点的；集中在点上的；病灶的，病灶性的
conical ['kɒnɪkl] *adj.* 圆锥（形）的
logo ['ləʊgəʊ] *n.* （某公司或机构的）标识，标志，徽标
apothecary [ə'pɒθəkərɪ] *n.* 药剂师，药店

Unit 7 Atomic Structure

The ancient Greek philosophers Leucippus and Democritus believed that atoms existed, but they had no idea as to their nature. Centuries later, in 1803, the English chemist John Dalton, guided by the experimental fact that chemical elements cannot be decomposed chemically, was led to formulate his atomic theory. Dalton's atomic theory was based on the assumption that atoms are tiny indivisible entities, with each chemical element consisting of its own characteristic atoms.

Fig. 7-1 Atomic structure.

The atom is now known to consist of three primary particles: protons, neutrons, and electrons, which make up the atoms of all matter. A series of experimental facts established the validity of the model. Radioactivity played an important part. Marie Curie suggested, in 1899, that when atoms disintegrate, they contradict Dalton's idea that atoms are indivisible. There must then be something smaller than the atom (subatomic particles) of which atoms were composed.

Long before that, Michael Faraday's electrolysis experiments and laws suggested that, just as an atom is the fundamental particle of an element, a fundamental particle for electricity must exist. The "particle" of electricity was given the name *electron*. Experiments with cathode-ray tubes, conducted by the British physicist Joseph John Thomson, proved the existence of the electron and obtained the charge-to-mass ratio for it. The experiments suggested that electrons are present in all kinds of matter and that they presumably exist in all atoms of all elements. Efforts were then turned to measuring the charge on the electron, and these were eventually successful by the American physicist Robert Andrews Millikan through the famous oil drop experiment.

The study of the so-called canal rays by the German physicist Eugen Goldstein, observed in a special cathode-ray tube with a perforated cathode, let to the recognition in 1902 that these rays were positively charged particles (*protons*). Finally, years later in 1932 the British physicist James Chadwick discovered another particle in the nucleus that had no charge, and for this reason was named neutron.

Fig. 7-2   British physicist and chemist John Dalton (1766—1844), by Charles Turner (1773—1857) after James Lonsdale (1777—1839). Mezzotint.

The idea that matter is made up of discrete units is a very old idea, appearing in many ancient cultures such as Greece and India. The word "atom" (Greek: ἄτομος; *atomos*), meaning "uncuttable", was coined by the Pre-Socratic Greek philosophers Leucippus and his pupil Democritus (460—370 BC). Democritus taught that atoms were infinite in number, uncreated, and eternal, and that the qualities of an object result from the kind of atoms that compose it. Democritus's atomism was refined and elaborated by the later Greek philosopher Epicurus (341—270 BC), and by the Roman Epicurean poet Lucretius (99—55 BC). During the Early Middle Ages, atomism was mostly forgotten in western Europe, but survived among some groups of Islamic philosophers. During the twelfth century, atomism became known again in western Europe through references to it in the newly-rediscovered writings of Aristotle.

In the fourteenth century, the rediscovery of major works describing atomist teachings, including Lucretius's *De rerum natura* and Diogenes Laërtius's *Lives and Opinions of Eminent Philosophers*, led to increased scholarly attention on the subject. Nonetheless, because atomism was associated with the philosophy of Epicureanism, which contradicted orthodox Christian teachings, belief in atoms was not considered acceptable by most European philosophers. The French Catholic priest Pierre Gassendi (1592—1655) revived Epicurean atomism with modifications, arguing that atoms were created by God and, though extremely numerous, are not infinite. Gassendi's modified theory of atoms was popularized in France by the physician François Bernier (1620—1688) and in England by the natural philosopher Walter Charleton (1619—1707). The chemist Robert Boyle (1627—1691) and the physicist Isaac Newton (1642—1727) both defended atomism and, by the end of the seventeenth century, it had become accepted by portions of the scientific community.

John Dalton

Near the end of the 18th century, two laws about chemical reactions emerged without referring to the notion of an atomic theory. The first was the law of conservation of mass, closely associated with the work of Antoine Lavoisier, which states that the total mass in a chemical reaction remains constant (that is, the reactants have the same mass as the products). The second was the law of definite proportions. First established by the French chemist Joseph Louis Proust in 1799, this law states that if a compound is broken down into its constituent chemical elements, then the masses of the constituents will always have the same proportions by weight, regardless of the quantity or source of the original substance.

John Dalton studied and expanded upon this previous work and defended a new idea, later known as the law of multiple proportions: if the same two elements can be combined to form a number of different compounds, then the ratios of the masses of the two elements in their various compounds will be represented by small whole numbers. For example, Proust had studied tin oxides and found that there is one type of tin oxide that is 88.1% tin and 11.9% oxygen and another type that is 78.7% tin and 21.3% oxygen, these are tin(II) oxide and tin dioxide respectively. Dalton noted from these percentages that 100g of tin will combine either with 13.5g or 27g of oxygen; 13.5 and 27 form a ratio of 1:2. Dalton found several examples of such instances of integral multiple combining proportions, and asserted that the pattern was a general one. Most importantly, he noted that an atomic theory of matter could elegantly explain this law, as well as Proust's law of definite proportions. For example, in the case of Proust's tin oxides, one tin atom will combine with either one or two oxygen atoms to form either the first or the second oxide of tin.

Dalton believed atomic theory could explain why water absorbed different gases in different proportions—for example, he found that water absorbed carbon dioxide far better than it absorbed nitrogen. Dalton hypothesized this was due to the differences in mass and complexity of the gases' respective particles. Indeed, carbon dioxide molecules (CO_2) are heavier and larger than nitrogen molecules (N_2).

Dalton proposed that each chemical element is composed of atoms of a single, unique type, and though they cannot be altered or destroyed by chemical means, they can combine to form more complex structures (chemical compounds). This marked the first truly scientific theory of the atom, since Dalton reached his conclusions by experimentation and examination of the results in an empirical fashion.

In 1803 Dalton orally presented his first list of relative atomic weights for a number of substances. This paper was published in 1805, but he did not discuss there exactly how he obtained these figures. The method was first revealed in 1807 by his acquaintance Thomas Thomson, in the third edition of Thomson's textbook, *A System of Chemistry*. Finally, Dalton published a full account in his own textbook, *A New System of Chemical Philosophy*, in 1808 and 1810.

Dalton estimated the atomic weights according to the mass ratios in which they combined, with the hydrogen atom taken as unity. However, Dalton did not conceive that with some elements atoms exist in molecules — e.g. pure oxygen exists as O_2. He also mistakenly believed that the simplest compound between any two elements is always one atom of each (so he thought water was HO, not H_2O). This, in addition to the crudity of his equipment, flawed his results. For instance, in 1803 he believed that oxygen atoms were 5.5 times heavier than hydrogen atoms, because in water he measured 5.5 grams of oxygen for every 1 gram of hydrogen and believed the formula for water was HO. Adopting better data, in 1806 he concluded that the atomic weight of oxygen must actually be 7 rather than 5.5, and he retained this weight for the rest of his life. Others at this time had already concluded that the oxygen atom must weigh 8 relative to hydrogen equals 1, if one assumes Dalton's formula for the water molecule (HO), or 16 if one assumes the modern water formula (H_2O).

New words

decompose [,diːkəm'pəʊz] *vt. & vi.* 分解；（使）腐烂
disintegrate [dɪs'ɪntɪɡreɪt] *vt.* 使某物碎裂，崩裂；使某物衰微，瓦解，分崩离析
　　　　　　　　　　　　　　vi. 碎裂，崩裂；衰微，瓦解，分崩离析
electrolysis [ɪ,lek'trɒləsɪs] *n.* 电解，电蚀；电蚀除瘤（毛发等）
cathode ray ['kæθ,əʊd reɪ] *n.* 阴极射线
canal ray [kə'næl reɪ] *n.* 极隧射线
perforate ['pɜːfəreɪt] *vt.* 穿孔于，在……上打眼
discrete [dɪ'skriːt] *adj.* 分离的，不相关联的；分立式；非连续
elaborate [ɪ'læbərət] *vi.* 详尽说明；变得复杂；
　　　　　　　　　　　vt. 详细制定；详尽阐述；[生理学]加工；尽心竭力地做
　　　　　　　　　　　adj. 复杂的；精心制作的；（结构）复杂的；精巧的
orthodox ['ɔːθədɒks] *adj.* 规范的；公认的；普遍赞同的；正统的
revive [rɪ'vaɪv] *vt.* 使复活，使恢复；使振奋，复原；使再生，使重新流行；唤醒，唤起
　　　　　　　　　 vi. 复苏，恢复；振作，恢复；再生，重新流行；再生效力

Unit 8   Chemical Bonding

Atoms sticking together in molecules or crystals are said to be bonded with one another. A chemical bond may be visualized as the multipole balance between the positive charges in the nuclei and the negative charges oscillating about them. More than simple attraction and repulsion, the energies and distributions characterize the availability of an electron to bond to another atom. These potentials create the interactions which hold atoms together in molecules or crystals. In many simple compounds, Valence Bond Theory, the Valence Shell Electron Pair Repulsion model (VSEPR), and the concept of oxidation number can be used to explain molecular structure and composition. Similarly, theories from classical physics can be used to predict many ionic structures. With more complicated compounds, such as metal complexes, valence bond theory fails and alternative approaches, primarily based on principles of quantum chemistry such as the molecular orbital theory, are necessary. See diagram on electronic orbitals.

A chemical bond is a lasting attraction between atoms, ions or molecules that enables the formation of chemical compounds. The bond may result from the electrostatic force of attraction between oppositely charged ions as in ionic bonds or through the sharing of electrons as in covalent bonds. The strength of chemical bonds varies considerably; there are "strong bonds" or "primary bonds" such as covalent, ionic and metallic bonds, and "weak bonds" or "secondary bonds" such as dipole–dipole interactions, the London dispersion force and hydrogen bonding.

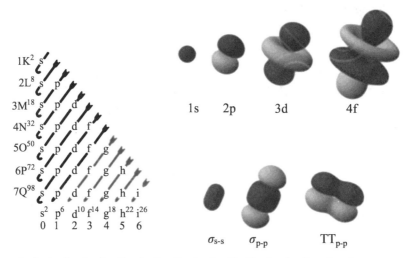

Fig. 8-1 Electron atomic and molecular orbitals.

Since opposite charges attract via a simple electromagnetic force, the negatively charged electrons that are orbiting the nucleus and the positively charged protons in the nucleus attract each other. An electron positioned between two nuclei will be attracted to both of them, and the nuclei will be attracted toward electrons in this position. This attraction constitutes the chemical bond. Due to the matter wave nature of electrons and their smaller mass, they must occupy a much larger amount of volume compared with the nuclei, and this volume occupied by the electrons keeps the atomic nuclei in a bond relatively far apart, as compared with the size of the nuclei themselves.

In general, strong chemical bonding is associated with the sharing or transfer of electrons between the participating atoms. The atoms in molecules, crystals, metals and diatomic gases—indeed most of the physical environment around us — are held together by chemical bonds, which dictate the structure and the bulk properties of matter.

All bonds can be explained by quantum theory, but, in practice, simplification rules allow chemists to predict the strength, directionality, and polarity of bonds. The octet rule and VSEPR theory are two examples. More sophisticated theories are valence bond theory which includes orbital hybridization and resonance, and molecular orbital theory which includes linear combination of atomic orbitals and ligand field theory. Electrostatics are used to describe bond polarities and the effects they have on chemical substances.

8.1 Overview of main types of chemical bonds

A chemical bond is an attraction between atoms. This attraction may be seen as the result of different behaviors of the outermost or valence electrons of atoms. These behaviors merge into each other seamlessly in various circumstances, so that there is no clear line to be drawn between them. However, it remains useful and customary to differentiate between different types of bond, which result in different properties of condensed matter.

In the simplest view of a covalent bond, one or more electrons (often a pair of electrons) are drawn into the space between the two atomic nuclei. Energy is released by bond formation. This is not as a reduction in potential energy, because the attraction of the two electrons to the two protons is offset by the electron-electron and proton-proton repulsions. Instead, the release of energy (and hence stability of the bond) arises from the reduction in kinetic energy due to the electrons being in a more spatially distributed (i.e. longer de Broglie wavelength) orbital compared with each electron being confined closer to its respective nucleus. These bonds exist between two particular identifiable atoms and have a direction in space, allowing them to be shown as single connecting lines between atoms in drawings, or modeled as sticks between spheres in models.

In a polar covalent bond, one or more electrons are unequally shared between two nuclei. Covalent bonds often result in the formation of small collections of better-connected atoms called molecules, which in solids and liquids are bound to other molecules by forces that are often much weaker than the covalent bonds that hold the molecules internally together. Such weak intermolecular bonds give organic molecular substances, such as waxes and oils, their soft bulk character, and their low melting points (in liquids, molecules must cease most structured or oriented contact with each other). When covalent bonds link long chains of atoms in large molecules, however (as in polymers such as nylon), or when covalent bonds extend in networks through solids that are not composed of discrete molecules (such as diamond or quartz or the silicate minerals in many types of rock) then the structures that result may be both strong and tough, at least in the direction oriented correctly with networks of covalent bonds. Also, the melting points of such covalent polymers and networks increase greatly.

In a simplified view of an ionic bond, the bonding electron is not shared at all, but transferred. In this type of bond, the outer atomic orbital of one atom has a vacancy which allows the addition of one or more electrons. These newly added electrons potentially occupy a lower energy-state (effectively closer to more nuclear charge) than they experience in a different atom. Thus, one nucleus offers a more tightly bound position to an electron than does another nucleus, with the result that one atom may transfer an electron to the other. This transfer causes one atom to assume a net positive charge, and the other to assume a net negative charge. The bond then results from electrostatic attraction between atoms and the atoms become positive or negatively charged ions. Ionic bonds may be seen as extreme examples of polarization in covalent bonds. Often, such bonds have no particular orientation in space, since they result from equal electrostatic attraction of each ion to all ions around them. Ionic bonds are strong (and thus ionic substances require high temperatures to melt) but also brittle, since the forces between ions are short-range and do not easily bridge cracks and fractures. This type of bond gives rise to the physical characteristics of crystals of classic mineral salts, such as table salt.

A less often mentioned type of bonding is metallic bonding. In this type of bonding, each atom in a metal donates one or more electrons to a "sea" of electrons that reside between many metal atoms. In this sea, each electron is free (by virtue of its wave nature) to be associated with a great many atoms at once. The bond results because the metal atoms become somewhat positively charged due to loss of their electrons while the electrons remain attracted to many atoms, without being part of any given atom. Metallic bonding may be seen as an extreme example of delocalization of electrons over a large system of covalent bonds, in which every atom participates. This type of bonding is often very strong (resulting in the tensile strength of metals). However, metallic bonding is more collective in nature than other types, and so they allow metal crystals to more easily deform, because they are composed of atoms attracted to each other, but not in any particularly-oriented ways. This results in the malleability of metals. The cloud of electrons in metallic bonding causes the characteristically good electrical and thermal conductivity of metals, and also their shiny lustre that reflects most frequencies of white light.

8.2 History

Early speculations about the nature of the chemical bond, from as early as the 12th century, supposed that certain types of chemical species were joined by a type of chemical affinity. In 1704,

Sir Isaac Newton famously outlined his atomic bonding theory, in "Query 31" of his Opticks, whereby atoms attach to each other by some "force". Specifically, after acknowledging the various popular theories in vogue at the time, of how atoms were reasoned to attach to each other, i.e. "hooked atoms", "glued together by rest", or "stuck together by conspiring motions", Newton states that he would rather infer from their cohesion, that "particles attract one another by some force, which in immediate contact is exceedingly strong, at small distances performs the chemical operations, and reaches not far from the particles with any sensible effect".

In 1819, on the heels of the invention of the voltaic pile, Jöns Jakob Berzelius developed a theory of chemical combination stressing the electronegative and electropositive characters of the combining atoms. By the mid19th century, Edward Frankland, F.A. Kekulé, A.S. Couper, Alexander Butlerov, and Hermann Kolbe, building on the theory of radicals, developed the theory of valency, originally called "combining power", in which compounds were joined owing to an attraction of positive and negative poles. In 1916, chemist Gilbert N. Lewis developed the concept of the electron-pair bond, in which two atoms may share one to six electrons, thus forming the single electron bond, a single bond, a double bond, or a triple bond; in Lewis's own words, "An electron may form a part of the shell of two different atoms and cannot be said to belong to either one exclusively".

That same year, Walther Kossel put forward a theory similar to Lewis' only his model assumed complete transfers of electrons between atoms, and was thus a model of ionic bonding. Both Lewis and Kossel structured their bonding models on that of Abegg's rule (1904).

Niels Bohr proposed a model of the atom and a model of the chemical bond. According to his model for a diatomic molecule, the electrons of the atoms of the molecule form a rotating ring whose plane is perpendicular to the axis of the molecule and equidistant from the atomic nuclei. The dynamic equilibrium of the molecular system is achieved through the balance of forces between the forces of attraction of nuclei to the plane of the ring of electrons and the forces of mutual repulsion of the nuclei. The Bohr model of the chemical bond took into account the Coulomb repulsion — the electrons in the ring are at the maximum distance from each other.

In 1927, the first mathematically complete quantum description of a simple chemical bond, i.e. that produced by one electron in the hydrogen molecular ion, H_2^+, was derived by the Danish physicist Oyvind Burrau. This work showed that the quantum approach to chemical bonds could be fundamentally and quantitatively correct, but the mathematical methods used could not be extended to molecules containing more than one electron. A more practical, albeit less quantitative, approach was put forward in the same year by Walter Heitler and Fritz London. The Heitler–London method forms the basis of what is now called valence bond theory. In 1929, the linear combination of atomic orbitals molecular orbital method (LCAO) approximation was introduced by Sir John Lennard-Jones, who also suggested methods to derive electronic structures of molecules of F_2 (fluorine) and O_2 (oxygen) molecules, from basic quantum principles. This molecular orbital theory represented a covalent bond as an orbital formed by combining the quantum mechanical Schrödinger atomic orbitals which had been hypothesized for electrons in single atoms. The equations for bonding electrons in multi-electron atoms could not be solved to mathematical perfection (i.e., analytically), but approximations for them still gave many good qualitative predictions and results. Most quantitative calculations in modern quantum chemistry use either valence bond or molecular orbital theory as a starting point, although a third approach, density functional theory, has become increasingly popular in recent years.

In 1933, H. H. James and A. S. Coolidge carried out a calculation on the dihydrogen molecule that, unlike all previous calculation which used functions only of the distance of the electron from the atomic nucleus, used functions which also explicitly added the distance between the two electrons. With up to 13 adjustable parameters they obtained a result very close to the experimental result for the dissociation energy. Later extensions have used up to 54 parameters and gave excellent agreement with experiments. This calculation convinced the scientific community that quantum theory could give agreement with experiment. However, this approach has none of the physical pictures of the valence bond and molecular orbital theories and is difficult to extend to larger molecules.

New words

oscillate ['ɒsɪleɪt] *vt.* 使振荡，使振动，使动摇
    vi. 持续周期性地摆动；动摇，犹豫
electrostatic [ɪˌlektrəʊ'stætɪk] *adj.* 静电的，静电学的；静电式
London dispersion force 伦敦力，伦敦色散力
directionality [dɪˌrekʃə'nælɪtɪ] *n.* 方向性，定向性；指向性
octet [ɒk'tet] *n.* 八位位组，八位字节
resonance ['rezənəns] *n.* 共鸣；反响；共振
ligand ['lɪgənd] *n.* 配合基，向心配合（价）体；配体
outermost ['aʊtəməʊst] *adj.* 最外面的，离中心最远的
seamless ['siːmləs] *adj.* 无缝的；无漏洞的
seamlessly *adv.* 无空隙地；无停顿地
covalent bond *n.* 共价键
brittle ['brɪtl] *adj.* 易碎的；难以相处的，尖刻暴躁的；冷淡的；声音尖利的
reside [rɪ'zaɪd] *vi.* 住，居住，（官吏）留驻；（性质）存在，具备，（权力等）属于，归于
deform [dɪ'fɔːm] *vt.* 使变形；使残废；使变丑；毁伤……的形体
    vi. 变形；变畸形
    adj. 畸形的；丑陋的
malleability [mələ'bɪlɪtɪ] *n.* 可锻性
lustre ['lʌstə(r)] *n.* 光泽；光荣；荣誉；出色
vogue [vəʊg] *n.* 时尚，流行；时髦的事物
    adj. 流行的，时髦的

conspire [kən'spaɪə] *vi.* 搞阴谋；协力促成
      vt. 阴谋策划
voltaic [vɒl'teɪɪk] *adj.* 电流的；伏打
radicals ['rædɪklz] *n.* 激进分子（radical 的名词复数）；根基；基本原理；[数学]根数
perpendicular [ˌpɜːpən'dɪkjələ(r)] *adj.* 垂直的，成直角的；直立的，险陡的
         n. 垂直线，垂直面；直立，直立姿势，廉直；垂直测器，锤规；[建]垂直式建筑，绝壁
axis ['æksɪs] *n.* 轴，轴线；[政]轴心；轴心国
equidistant [ˌiːkwɪ'dɪstənt] *adj.* 距离相等的，等距的
albeit [ˌɔːl'bɪːɪt] *conj.* 虽然；即使
dihydrogen [diː'haɪdrədʒən] *adj.* & *n.* [医]二氢化的，二氢

Unit 9　Chemical Reaction

When a chemical substance is transformed as a result of its interaction with another or energy, a chemical reaction is said to have occurred. Chemical reaction is a therefore a concept related to the reaction of a substance when it comes in close contact with another, whether as a mixture or a solution; or its exposure to a some form of energy. It results in some energy exchange between the constituents of the reaction as well with the system environment which may be a designed vessels which are often laboratory glassware. Chemical reactions can result in the formation or dissociation of molecules, that is, molecules breaking apart to form two or more smaller molecules, or rearrangement of atoms within or across molecules. Chemical reactions usually involve the making or breaking of chemical bonds. Oxidation, reduction, dissociation, acid-base neutralization and molecular rearrangement are some of the commonly used kinds of chemical reactions.

A chemical reaction is a transformation of some substances into one or more other substances. It can be symbolically depicted through a chemical equation. The number of atoms on the left and the right in the equation for a chemical transformation is most often equal. The nature of chemical reactions a substance may undergo and the energy changes that may accompany it are constrained by certain basic rules, known as chemical laws.

Energy and entropy considerations are invariably important in almost all chemical studies. Chemical substances are classified in terms of their structure, phase as well as their chemical compositions. They can be analyzed using the tools of chemical analysis, e.g. spectroscopy and chromatography.

While in a non-nuclear chemical reaction the number and kind of atoms on both sides of the equation are equal, for a nuclear reaction this holds true only for the nuclear particles viz. protons and neutrons.

The sequence of steps in which the reorganization of chemical bonds may be taking place in the course of a chemical reaction is called its mechanism. A chemical reaction can be envisioned to take place in a number of steps, each of which may have a different speed. Many reaction intermediates with variable stability can thus be envisaged during the course of a reaction. Reaction mechanisms are proposed to explain the kinetics and the relative product mix of a reaction. Many physical chemists specialize in exploring and proposing the mechanisms of various chemical reactions. Several empirical rules, like the Woodward-Hoffmann rules often come handy while proposing a mechanism for a chemical reaction.

According to the IUPAC gold book a chemical reaction is a process that results in the interconversion of chemical species. Accordingly, a chemical reaction may be an elementary reaction or a stepwise reaction. An additional caveat is made, in that this definition includes cases where the interconversion of conformers is experimentally observable. Such detectable chemical reactions normally involve sets of molecular entities as indicated by this definition, but it is often conceptually convenient to use the term also for changes involving single molecular entities (i.e. "microscopic chemical events").

Four basic types

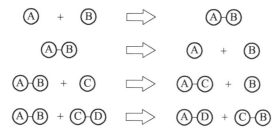

Fig. 9-1 The four basic chemical reactions types: synthesis, decomposition, single replacement and double replacement.

9.1 Synthesis

In a synthesis reaction, two or more simple substances combine to form a more complex substance.

$$A + B \longrightarrow AB$$

"Two or more reactants giving one product" is another way to identify a synthesis reaction. One example of a synthesis reaction is the combination of iron and sulfur to form iron(II) sulfide:

$$8Fe + S_8 \longrightarrow 8FeS$$

Another example is simple hydrogen gas combined with simple oxygen gas to produce a more complex substance, such as water.

9.2 Decomposition

A decomposition reaction is when a more complex substance breaks down into its more simple parts. It is thus the opposite of a synthesis reaction, and can be written as:

$$AB \longrightarrow A + B$$

One example of a decomposition reaction is the electrolysis of water to make oxygen and hydrogen gas:

$$2H_2O \longrightarrow 2H_2 + O_2$$

9.3 Single replacement

In a single replacement reaction, a single uncombined element replaces another in a compound;

in other words, one element trades places with another element in a compound. These reactions come in the general form of:

$$A + BC \longrightarrow AC + B$$

One example of a single displacement reaction is when magnesium replaces hydrogen in water to make magnesium hydroxide and hydrogen gas:

$$Mg + 2H_2O \longrightarrow Mg(OH)_2 + H_2$$

9.4 Double replacement

In a double replacement reaction, the anions and cations of two compounds switch places and form two entirely different compounds. These reactions are in the general form:

$$AB + CD \longrightarrow AD + CB$$

For example, when barium chloride ($BaCl_2$) and magnesium sulfate ($MgSO_4$) react, the SO_4^{2-} anion switches places with the $2Cl^-$ anion, giving the compounds $BaSO_4$ and $MgCl_2$.

Another example of a double displacement reaction is the reaction of lead(II) nitrate with potassium iodide to form lead(II) iodide and potassium nitrate:

$$Pb(NO_3)_2 + 2KI \longrightarrow PbI_2 + 2KNO_3$$

Chemical reaction, a process in which one or more substances, the reactants, are converted to one or more different substances, the products. Substances are either chemical elements or compounds. A chemical reaction rearranges the constituent atoms of the reactants to create different substances as products.

Chemical reactions are an integral part of technology, of culture, and indeed of life itself. Burning fuels, smelting iron, making glass and pottery, brewing beer, and making wine and cheese are among many examples of activities incorporating chemical reactions that have been known and used for thousands of years. Chemical reactions abound in the geology of Earth, in the atmosphere and oceans, and in a vast array of complicated processes that occur in all living systems.

Chemical reactions must be distinguished from physical changes. Physical changes include changes of state, such as ice melting to water and water evaporating to vapour. If a physical change occurs, the physical properties of a substance will change, but its chemical identity will remain the same. No matter what its physical state, water (H_2O) is the same compound, with each molecule composed of two atoms of hydrogen and one atom of oxygen. However, if water, as ice, liquid, or vapour, encounters sodium metal (Na), the atoms will be redistributed to give the new substances molecular hydrogen (H_2) and sodium hydroxide (NaOH). By this, we know that a chemical change or reaction has occurred.

9.5 Historical overview

The concept of a chemical reaction dates back about 250 years. It had its origins in early experiments that classified substances as elements and compounds and in theories that explained these processes. Development of the concept of a chemical reaction had a primary role in defining the science of chemistry as it is known today.

The first substantive studies in this area were on gases. The identification of oxygen in the 18th century by Swedish chemist Carl Wilhelm Scheele and English clergyman Joseph Priestley had particular significance. The influence of French chemist Antoine-Laurent Lavoisier was especially notable, in that his insights confirmed the importance of quantitative measurements of chemical processes. In his book *Traité élémentaire de chimie* (1789; *Elementary Treatise on Chemistry*), Lavoisier identified 33 "elements" — substances not broken down into simpler entities. Among his many discoveries, Lavoisier accurately measured the weight gained when elements were oxidized, and he ascribed the result to the combining of the element with oxygen. The concept of chemical reactions involving the combination of elements clearly emerged from his writing, and his approach led others to pursue experimental chemistry as a quantitative science.

The other occurrence of historical significance concerning chemical reactions was the development of atomic theory. For this, much credit goes to English chemist John Dalton, who postulated his atomic theory early in the 19th century. Dalton maintained that matter is composed of small, indivisible particles, that the particles, or atoms, of each element were unique, and that chemical reactions were involved in rearranging atoms to form new substances. This view of chemical reactions accurately defines the current subject. Dalton's theory provided a basis for understanding the results of earlier experimentalists, including the law of conservation of matter (matter is neither created nor destroyed) and the law of constant composition (all samples of a substance have identical elemental compositions).

Thus, experiment and theory, the two cornerstones of chemical science in the modern world, together defined the concept of chemical reactions. Today experimental chemistry provides innumerable examples, and theoretical chemistry allows an understanding of their meaning.

9.6 Basic concepts of chemical reactions

9.6.1 Synthesis

When making a new substance from other substances, chemists say either that they carry out a synthesis or that they synthesize the new material. Reactants are converted to products, and the process is symbolized by a chemical equation. For example, iron (Fe) and sulfur (S) combine to form iron sulfide (FeS): Fe(s) + S(s) → FeS(s). The plus sign indicates that iron reacts with sulfur. The arrow signifies that the reaction "forms" or "yields" iron sulfide, the product. The state of matter of reactants and products is designated with the symbols (s) for solids, (l) for liquids, and (g) for gases.

9.6.2 The conservation of matter

In reactions under normal laboratory conditions, matter is neither created nor destroyed, and elements are not transformed into other elements. Therefore, equations depicting reactions must be balanced; that is, the same number of atoms of each kind must appear on opposite sides of the equation. The balanced equation for the iron-sulfur reaction shows that one iron atom can react with one sulfur atom to give one formula unit of iron sulfide.

Chemists ordinarily work with weighable quantities of elements and compounds. For example, in the iron-sulfur equation the symbol Fe represents 55.845 grams of iron, S represents 32.066 grams of sulfur, and FeS represents 87.911 grams of iron sulfide. Because matter is not created or destroyed in a chemical reaction, the total mass of reactants is the same as the total mass of products. If some other amount of iron is used, say, one-tenth as much (5.585 grams), only one-tenth as much sulfur can be consumed (3.207 grams), and only one-tenth as much iron sulfide is produced (8.791 grams). If 32.066 grams of sulfur were initially present with 5.585 grams of iron, then 28.859 grams of sulfur would be left over when the reaction was complete.

The reaction of methane (CH_4, a major component of natural gas) with molecular oxygen (O_2) to produce carbon dioxide (CO_2) and water can be depicted by the chemical equation:

$$CH_4(g) + 2O_2(g) \longrightarrow CO_2(g) + 2H_2O(l)$$

Here another feature of chemical equations appears. The number 2 preceding O_2 and H_2O is a stoichiometric factor. The number 1 preceding CH_4 and CO_2 is implied. This indicates that one molecule of methane reacts with two molecules of oxygen to produce one molecule of carbon dioxide and two molecules of water. The equation is balanced because the same number of atoms of each element appears on both sides of the equation (here one carbon, four hydrogen, and four oxygen atoms). Analogously with the iron-sulfur example, we can say that 16 grams of methane and 64 grams of oxygen will produce 44 grams of carbon dioxide and 36 grams of water. That is, 80 grams of reactants will lead to 80 grams of products.

The ratio of reactants and products in a chemical reaction is called chemical stoichiometry. Stoichiometry depends on the fact that matter is conserved in chemical processes, and calculations giving mass relationships are based on the concept of the mole. One mole of any element or compound contains the same number of atoms or molecules, respectively, as one mole of any other element or compound. By international agreement, one mole of the most common isotope of carbon (carbon-12) has a mass of exactly 12 grams (this is called the molar mass) and represents $6.022140857 \times 10^{23}$ atoms (Avogadro's number). One mole of iron contains 55.847 grams; one mole of methane contains 16.043 grams; one mole of molecular oxygen is equivalent to 31.999 grams; and one mole of water is 18.015 grams. Each of these masses represents $6.022140857 \times 10^{23}$ molecules.

9.6.3 Energy considerations

Energy plays a key role in chemical processes. According to the modern view of chemical reactions, bonds between atoms in the reactants must be broken, and the atoms or pieces of molecules are reassembled into products by forming new bonds. Energy is absorbed to break bonds, and energy is evolved as bonds are made. In some reactions the energy required to break bonds is larger than the energy evolved on making new bonds, and the net result is the absorption of energy. Such a reaction is said to be endothermic if the energy is in the form of heat. The opposite of endothermic is exothermic; in an exothermic reaction, energy as heat is evolved. The more general terms *exoergic* (energy evolved) and *endoergic* (energy required) are used when forms of energy other than heat are involved.

A great many common reactions are exothermic. The formation of compounds from the constituent elements is almost always exothermic. Formation of water from molecular hydrogen and oxygen and the formation of a metal oxide such as calcium oxide (CaO) from calcium metal and oxygen gas are examples. Among widely recognizable exothermic reactions is the combustion of fuels (such as the reaction of methane with oxygen mentioned previously).

The formation of slaked lime (calcium hydroxide, $Ca(OH)_2$) when water is added to lime (CaO) is exothermic. $CaO(s) + H2O (l) \longrightarrow Ca(OH)_2(s)$ This reaction occurs when water is added to dry portland cement to make concrete, and heat evolution of energy as heat is evident because the mixture becomes warm.

Not all reactions are exothermic (or exoergic). A few compounds, such as nitric oxide (NO) and hydrazine (N_2H_4), require energy input when they are formed from the elements. The decomposition of limestone ($CaCO_3$) to make lime (CaO) is also an endothermic process; it is necessary to heat limestone to a high temperature for this reaction to occur. $CaCO_3(s) \longrightarrow CaO(s) + CO_2(g)$

The decomposition of water into its elements by the process of electrolysis is another endoergic process. Electrical energy is used rather than heat energy to carry out this reaction $2H_2O(g) \longrightarrow 2 H_2(g) + O_2(g)$. Generally, evolution of heat in a reaction favours the conversion of reactants to products. However, entropy is important in determining the favourability of a reaction. Entropy is a measure of the number of ways in which energy can be distributed in any system. Entropy accounts for the fact that not all energy available in a process can be manipulated to do work.

A chemical reaction will favour the formation of products if the sum of the changes in entropy for the reaction system and its surroundings is positive. An example is burning wood. Wood has a low entropy. When wood burns, it produces ash as well as the high-entropy substances carbon dioxide gas and water vapour. The entropy of the reacting system increases during combustion. Just as important, the heat energy transferred by the combustion to its surroundings increases the entropy in the surroundings. The total of entropy changes for the substances in the reaction and the surroundings is positive, and the reaction is product-favoured.

When hydrogen and oxygen react to form water, the entropy of the products is less than that of the reactants. Offsetting this decrease in entropy, however, is the increase in entropy of the surroundings owing to the heat transferred to it by the exothermic reaction. Again because of the overall increase in entropy, the combustion of hydrogen is product-favoured.

9.6.4 Kinetic considerations

Chemical reactions commonly need an initial input of energy to begin the process. Although the combustion of wood, paper, or methane is an exothermic process, a burning match or a spark is needed to initiate this reaction. The energy supplied by a match arises from an exothermic chemical reaction that is itself initiated by the frictional heat generated by rubbing the match on a suitable surface.

In some reactions, the energy to initiate a reaction can be provided by light. Numerous reactions in Earth's atmosphere are photochemical, or light-driven, reactions initiated by solar radiation. One example is the transformation of ozone (O_3) into oxygen (O_2) in the troposphere. The absorption of ultraviolet light (hv) from the Sun to initiate this reaction prevents potentially harmful high-energy radiation from reaching Earth's surface.

For a reaction to occur, it is not sufficient that it be energetically product-favoured. The reaction must also occur at an observable rate. Several factors influence reaction rates, including the concentrations of reactants, the temperature, and the presence of catalysts. The concentration affects the rate at which reacting molecules collide, a prerequisite for any reaction. Temperature is influential because reactions occur only if collisions between reactant molecules are sufficiently energetic. The proportion of molecules with sufficient energy to react is related to the temperature. Catalysts affect rates by providing a lower energy pathway by which a reaction can occur. Among common catalysts are precious metal compounds used in automotive exhaust systems that accelerate the breakdown of pollutants such as nitrogen dioxide into harmless nitrogen and oxygen. A wide array of biochemical catalysts are also known, including chlorophyll in plants (which facilitates the reaction by which atmospheric carbon dioxide is converted to complex organic molecules such as glucose) and many biochemical catalysts called enzymes. The enzyme pepsin, for example, assists in the breakup of large protein molecules during digestion.

New words

dissociation [dɪˌsəʊsɪ'eɪʃn] *n.* 分离；离解；脱离关系；解体
chromatography [ˌkrəʊmə'tɒgrəfɪ] *n.* 套色版；层析法
stepwise ['stepwaɪz] *adj.* 楼梯式的，逐步的
caveat ['kævɪæt] *n.* 警告，附加说明
magnesium [mæg'niːzɪəm] *n.* [化]镁
anion ['ænaɪən] *n.* 阴离子
cation ['kætaɪən] *n.* 阳离子
iodide ['aɪədaɪd] *n.* 碘化物
substantive [səb'stæntɪv] *adj.* 真实的；独立的；大量的；本质的，实质的
                n. 作名词用的词或词组；[语]实词，名词；独立存在的实体

postulate ['pɒstjuleɪt] *vt.* 假定；提出要求；视……为理所当然
　　　　　　n. 假定；先决条件；基本原理
methane ['mi:θeɪn] *n.* <化>甲烷，沼气
stoichiometric [stɔɪkɪ'ɒmɪtrɪk] *adj.* 化学当量的，化学计算的
analogously [ə'næləgəslɪ] *adv.* 类似地，近似地
stoichiometry [ˌstɔɪkɪ'ɒmɪtrɪ] *n.* 化学计算（法），化学计量学
net [net] *n.* 网；网状织物；球网；网罩
　　　vt. 用网捕；捕获；净赚；踢入球门
　　　adj. 净的；净得的；最后的
endothermic [ˌendəʊ'θɜːmɪk] *adj.* <化>吸热的；吸能的；由吸热而产生的；<动>温血的
exothermic [ˌeksəʊ'θɜːmɪk] *adj.* 发热的，放出热量的
exoergic [eksəʊ'ədʒɪk] *adj.* 放能的
endoergic [endəʊ'ɜːdʒɪk] *adj.* 吸能的，吸热的
slake [sleɪk] *vt.* 解（渴）；消除
lime [laɪm] *n.* 酸橙；石灰；绿黄色；椴树
　　　vt. 撒石灰于；涂粘鸟胶于
offset ['ɒfset] *vt.* 抵消；补偿；（为了比较的目的而）把……并列（或并置）
　　　vi. 形成分支，长出分枝；装支管
　　　n. 开端；出发；平版印刷；抵消，补偿
　　　adj. 分支的；偏（离中）心的；抵消的；开端的
kinetic [kɪ'netɪk] *adj.* 运动的，活跃的，能动的，有力的；[物]动力（学）的，运动的
ozone ['əʊzəʊn] *n.* [化]臭氧；清新空气
troposphere ['trɒpəsfɪə(r)] *n.* 对流层
ultraviolet [ˌʌltrə'vaɪələt] *adj.* 紫外的；紫外线的；产生紫外线的
　　　　　　n. 紫外线辐射；紫外光
catalyst ['kætəlɪst] *n.* <化>触媒，催化剂；〈比喻〉触发因素；促进因素
collide [kə'laɪd] *vi.* 相撞；碰撞；冲突；抵触
prerequisite [ˌpriː'rekwəzɪt] *n.* 先决条件，前提，必要条件
　　　　　　adj. 必须先具备的，必要的；先决条件的
automotive [ˌɔːtə'məʊtɪv] *adj.* 自动的；汽车的
chlorophyll ['klɒrəfɪl] *n.* 叶绿素
enzyme ['enzaɪm] *n.* [生化]酶
pepsin ['pepsɪn] *n.* 胃蛋白酶

Unit 10　Factors Influencing the Rate of a Chemical Reaction

It's useful to be able to predict whether an action will affect the rate at which a chemical reaction proceeds. There are several factors that can influence chemical reaction rate. In general, a factor that increases the number of collisions between particles will increase the reaction rate and a factor that decreases the number of collisions between particles will decrease the chemical reaction rate.

10.1　Concentration of reactants

A higher concentration of reactants leads to more effective collisions per unit time, which leads to an increased reaction rate (except for zero order reactions). Similarly, a higher concentration of products tends to be associated with a lower reaction rate. Use the partial pressure of reactants in a gaseous state as a measure of their concentration.

10.2　Temperature

Usually, an increase in temperature is accompanied by an increase in the reaction rate. Temperature is a measure of the kinetic energy of a system, so higher temperature implies higher average kinetic energy of molecules and more collisions per unit time. A general rule of thumb for most (not all) chemical reactions is that the rate at which the reaction proceeds will approximately double for each 10°C increase in temperature. Once the temperature reaches a certain point, some of the chemical species may be altered (e.g., denaturing of proteins) and the chemical reaction will slow or stop.

10.3　Medium or state of matter

The rate of a chemical reaction depends on the medium in which the reaction occurs. It may make a difference whether a medium is aqueous or organic; polar or nonpolar; or liquid, solid, or gaseous. Reactions involving liquids and especially solids depend on the available surface area.

For solids, the shape and size of the reactants make a big difference in the reaction rate.

10.4　Presence of catalysts and competitors

Catalysts (e.g., enzymes) lower the activation energy of a chemical reaction and increase the rate of a chemical reaction without being consumed in the process. Catalysts work by increasing the frequency of collisions between reactants, altering the orientation of reactants so that more collisions are effective, reducing intramolecular bonding within reactant molecules, or donating electron density to the reactants. The presence of a catalyst helps a reaction to proceed more quickly to equilibrium. Aside from catalysts, other chemical species can affect a reaction. The quantity of hydrogen ions (the pH of aqueous solutions) can alter a reaction rate. Other chemical species may compete for a reactant or alter orientation, bonding, electron density, etc., thereby decreasing the rate of a reaction.

10.5 Pressure

Increasing the pressure of a reaction improves the likelihood reactants will interact with each other, thus increases the rate of the reaction. As you would expect, this factor is important for reactions involving gases, and not a significant factor with liquids and solids.

10.6 Mixing

Mixing reactants together increases their ability to interact, thus increases the rate of a chemical reaction.

10.7 Summary of factors that affect chemical reaction rate

The chart below is a summary of the main factors that influence reaction rate. Keep in mind, there is typically a maximum effect, after which changing a factor will have no effect or will slow a reaction. For example, increasing temperature past a certain point may denature reactants or cause them to undergo a completely different chemical reaction.

Table 10-1 Factors that affect chemical reaction rate

Factor	Affect on Reaction Rate
temperature	increasing temperature increases reaction rate
pressure	increasing pressure increases reaction rate
concentration	in a solution, increasing the amount of reactants increases the reaction rate
state of matter	gases react more readily than liquids, which react more readily than solids
catalysts	a catalyst lowers activation energy, increasing reaction rate
mixing	mixing reactants improves reaction rate

New words

thumb [θʌm] *n.* 拇指；（手套的）拇指部分
  vi. 伸出拇指请求搭乘（过路汽车），示意请求搭便车；翻阅
  vt. 翻阅；作搭车手势；用拇指翻脏（书页等）
denature [dɪˈneɪtʃə] *vt.* 使改变本性；使变质；使中毒
aqueous [ˈeɪkwɪəs] *adj.* 水的，水成的
intramolecular [ˌɪntrəməˈlekjʊlə] *adj.* 作用（存在，发生）于分子内的

Part 3

Laboratory

Unit 11 Laboratory Apparatus: Chemical Instruments

Fig. 11-1 test tubes
试管

Fig. 11-2 test tube brushes
试管刷

Fig. 11-3 test tube holder
试管夹

Fig. 11-4 test tube rack
试管架

Fig. 11-5 beaker
烧杯

Fig. 11-6 stirring rods
搅拌棒

Fig. 11-7 thermometer
温度计

Fig. 11-8 boiling flask
长颈烧瓶

Fig. 11-9 Florence flask
平底烧瓶

Fig. 11-10 flask, round bottom, two-neck
双口圆底烧瓶

Fig. 11-11 boling flask, three-neck
三口圆底烧瓶

Fig. 11-12 flask, round bottom, four-neck
四口圆底烧瓶

Part 3 Laboratory 95

Fig. 11-13 conical flask; Erlenmeyer flask
锥形烧瓶，爱伦美氏（烧）瓶

Fig. 11-14 wide-mouth bottle
广口瓶

Fig. 11-15 graduated cylinder
量筒

Fig. 11-16 gas measuring tube
气体测量管

Fig. 11-17 volumetric flask
（容）量瓶

Fig. 11-18 geiser burette (stopcock)
酸氏滴定管

Fig. 11-19 transfer pipette
移液管；移液吸管

Fig. 11-20 ground joint
磨口连接

Fig. 11-21 mohr burette (with pinchcock)
莫尔滴定管

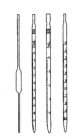

Fig. 11-22 Mohr measuring pipette
量液管

Fig. 11-23 watch glass
表面皿

Fig. 11-24 evaporating dish
蒸发皿

Fig. 11-25 petri dish
培养皿

Fig. 11-26　funnel
漏斗

Fig. 11-27　long-stem funnel
长颈漏斗

Fig. 11-28　filter funnel
过滤漏斗

Fig. 11-29　Büchner funnel
布氏漏斗

Fig. 11-30　separatory funnel
分液漏斗

Fig. 11-31　Hirsch funnel
赫希漏斗

Fig. 11-32　filter flask
抽滤瓶

Fig. 11-33　thiele melting point tube
熔点管（B 形管）

Fig. 11-34　plastic squeeze bottle (wash bottle)
洗瓶

Fig. 11-35　medicine dropper
胶头滴管

Fig. 11-36　rubber pipette bulb
洗耳球

Fig. 11-37　pipette (pipet)
移液管

Part 3   Laboratory   97

Fig. 11-38   chemical spoon
药勺

Fig. 11-39   microspatula
微量药勺

Fig. 11-40   mortar and pestle
研钵和研杵

Fig. 11-41   filter paper
滤纸

Fig. 11-42   Bunsen burner
本生灯

Fig. 11-43   burette stand
滴定管架（铁架台）

Fig. 11-44   extension support ring
延长支撑环

Fig. 11-45   ring stand (with ring)
铁架吕（带支撑环）

Fig. 11-46   crucible
坩埚

Fig. 11-47   crucible tong
坩埚钳

Fig. 11-48   beaker tong
烧杯钳

Fig. 11-49   universal extension
       clamp
普通延长夹

Fig. 11-50　3 prong clamp
三爪夹

Fig. 11-51　utility clamp
铁试器夹

Fig. 11-52　burette clamp
滴定管夹

Fig. 11-53　hose clamp
软管夹

Fig. 11-54　pinchcock; pinch clamp
弹簧夹

Fig. 11-55　screw clamp
螺旋夹

Fig. 11-56　ring clamp
环夹

Fig. 11-57　desiccator
干燥器

Fig. 11-58　goggle
护目镜

Fig. 11-59　stopcock
旋塞阀

Fig. 11-60　wire gauze
金属丝网（石棉网）

Fig. 11-61　centrifuge tube
离心管

Part 3 Laboratory   99

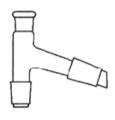

Fig. 11-62　distilling head
蒸馏头

Fig. 11-63　distilling tube
分馏管

Fig. 11-64　distilling tube, three bulbs
三球分馏管

Fig. 11-65　side-arm distillation flask
支管蒸馏瓶

Fig. 11-66　Claisen distilling head
减压蒸馏头

Fig. 11-67　air condenser
空气冷凝器

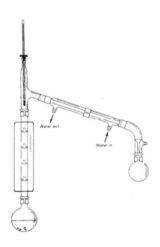

Fig. 11-68　fractionating column
分馏装置

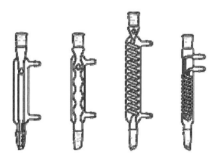

Fig. 11-69　condensers
凝结器

Unit 12 Recrystallization

Laboratory techniques are the set of procedures used on natural sciences such as chemistry, biology, physics to conduct an experiment, all of them follow the scientific method; while some of them involve the use of complex laboratory equipment from laboratory glassware to electrical devices, and others require more specific or expensive supplies.

Seldom do we encounter pure materials. Instead, many materials are mixtures made up of two or more chemically different substances. In order to isolate pure components of a mixture, chemists have developed a variety of techniques for the separation of one component from another, taking advantage of the differences in physical properties of the components. Recrystallization is one of the important laboratory processes frequently used for this.

Recrystallization is used to purify a solid substance at the temperature of the experiment. It is a basic purification technique based on different solubilities of solids. Insoluble impurities can be easily removed by filtration after dissolution of the solid that needs to be purified, while small amount of soluble impurities remains in the solution. Increasing the temperature produces a supersaturated solution which can be used to obtain crystals of the pure solid. When slowly cooling the solution down to room temperature, crystals form and crash out, with the impurities in the solution. Sometimes it is easier to conduct recrystallization using tow solvents, one good solvent for the compound and one poor solvent.

12.1 Single-solvent approach

A single-solvent recrystallization includes the following steps: selecting the solvent; dissolving the solid; cooling the solution; filtering and drying the crystals.

12.1.1 Selecting the solvent

Choosing an appropriate solvent is the first step in a recrystallization. Water, hexane, methanol, and ethyl acetate are frequently used. Ideally, the solid is virtually insoluble in the solvent at room temperature, yet is completely soluble at higher temperatures at or near the boiling point of the solvent.

To find a suitable solvent, it is necessary to test the solubility of the desired compound in different solvents. Test tubes and a rack, a test tube clamp, pipets and bulbs, a spatula, a beaker, and a hot plate are required, in addition to the compound, water, and the solvents.

Load a small amount of solid into a test tube, followed by adding about one milliliter of the test solvent. If the solid dissolves immediately at room temperature, the solvent is not suitable for recrystallization. Repeat this process with another test solvent using a clean test tube. If the solid does not dissolve, heat up the test tube using a hot water bath whose temperature is set at the boiling point of the test solvent. If the solid still remains, then this solvent is not good, either. Repeat with other solvents until the solid remains at room temperature, but dissolves in the solvent with the temperature at the boiling point, indicating a good recrystallization solvent.

12.1.2 Dissolving the solid

In the second step, the solid to be recrystallized is dissolved in the hot suitable solvent. Two Erlenmeyer flasks (one for the solvent and the other for the crystals), a hot plate, a disposable pipet and bulb, finger cots, and some boiling stones are needed in this step. Place two boiling stones in each flask to ensure smooth boiling during heating. A small amount of solvent is added to a flask containing the impure solid, and then the suspension in the flask is heated to the boiling point of the solvent until the complete dissolution of the solid. If the solid does not dissolve, add more hot solvent drop-wise continually until the solid is fully dissolved. A hot filtration is required if the solution contains visible solid impurities other than boiling stones. If the solution appears colored, the hot saturated solution is boiled for a short period of time with the addition of activated carbon to remove colored impurities, followed by a hot filtration to get rid of the activated carbon.

12.1.3 Cooling the solution

Next, the solution is cooled for the desired compound to crystallize. A more pure solid precipitates out from the solution, leaving soluble impurities in the solvent. The Nobel laureate, the late Professor Robert Bums Woodward stated that crystallization is one of the most beautiful processes known, and no true chemist fails to experience a thrill when he brings a new form of matter into the crystalline state for the first time.

In most cases, crystals grow as the solution cools down. Leave the solution undisturbed until the temperature decreases, and crystals begin to form on the bottom of the flask. Usually slower cooling leads to a more pure product. The size of crystals that form also depends on the cooling rate. Very small crystals tend to form upon rapid cooling and the impurities may also precipitate out of the solution along with the small crystals. Therefore, it is quite common to allow the solution to cool to room temperature first before cooling it further by setting the flask in an ice-water bath. Wait for the majority of crystals to form at room temperature and then place the flask in an ice-water bath.

However, sometimes crystallization needs to be induced by nucleation. One method is to scratch the flask with a glass rod at the air-solvent meniscus. The scratch increases the surface area of the glass, resulting in a roughened surface on which the solid can nucleate and crystallize. Another technique is to add a small crystal of the desired pure solid as the "seed" into the cooled solution if such a crystal is available. The "seed" crystal serves as the nucleating site for the crystal to grow.

Make sure that the solution is cool; otherwise, the added small crystal would dissolve.

If there are no crystals falling out of the solution, it is possible that too much solvent has been used. The solution should be concentrated further by allowing some of the solvent to evaporate. If crystals do not immediately form, reheat and then cool the solution.

12.1.4　Filtering and drying the crystals

After crystals have formed, it is time to separate them from the solution. Vacuum filtration is frequently used to isolate and dry the purified solid, sometimes washing the purified solid with chilled solvent. Use the smallest possible amount of cold solvent when washing the product to avoid dissolving some of the sample.

In the apparatus, vacuum is supplied by a pump and applied to the filter flask through a rubber tubing. Add filter paper to the funnel which is placed on the filter vacuum adapter in the neck of the filter flask. Use a small amount of the recrystallization solvent to moisten the filter paper and then turn on the pump. Pour and transfer the crystals and solution to the center of the filter paper. Add cold solvent to the flask and swirl the remaining crystals into the funnel.

Once the liquid is all sucked through, turn off the pump to release the vacuum. Then add a small amount of cold, clean solvent to wash the crystals and apply a gentle suction to allow the fresh solvent passing through the crystals at a slower rate. Note that suction should not be applied while washing. In order to dry the crystals as thoroughly as possible, full suction is applied for a few minutes. Drying the product via vacuum filtration should remove much of the solvent. Depending on the volatility of the solvent, sometimes open-air drying is used as well.

After filtering and drying, the final step is to remove the crystals from the filter funnel. Use a spatula to transfer the crystals to a watch glass. Physically separate any remaining boiling stones from the crystals in this step. In some cases, the recrystallization process is repeated to further purify the substance.

12.2　Two-solvent recrystallization

A two-solvent recrystallization includes the following steps: selecting the solvents; dissolving the solid; cooling the solution; filtering and drying the crystals.

12.2.1　Selecting the solvents

Similar to the single-solvent approach, the first step is also to select suitable solvents. As mentioned above, two solvents are needed, with one being a very good solvent for the compound and the other extremely poor at room temperature. And these two solvents must be miscible.

A glass plate, a spatula, and several clean Pasteur pipets and bulbs are needed in this step along with a range of candidate solvents and the compound to be purified. To select the solvents, only a small amount of solid compound is needed. On a glass plate a tiny amount of the compound is placed then about four centimeters away another sample is added. In a similar fashion place more solid samples on the glass plate until there are enough samples for the number of solvents to be chosen from. Select solvents with different polarity such as water, methanol, ethyl acetate, and hexane. Take three or four drops of one test solvent and add them to one solid sample. And repeat this process for the remaining solvents. Check the solubility results and evaluate whether the compound dissolves completely, partially, or not at all. Again, the perfect combination of solvents means that one solvent (Solvent A) easily dissolves the compound and the second solvent (Solvent B) does not dissolve the compound.

12.2.2 Dissolving the solid

In this recrystallization approach, the two solvents A and B should also be hot. Add each solvent in an Erlenmeyer flask along with boiling stones. Then heat up the solvents until near their respective boiling points. Load the impure compound in a tared test tube that is no more than one quarter full of solid.

Add the first recrystallization Solvent A, to dissolve the crystals. Add just enough hot Solvent A with a Pasteur pipet to the test tube that contains the compound. During additions of Solvent A, heat and shake the test tube to help dissolve the compound. Minimum amount of hot Solvent A should be used and the volume of Solvent A should not exceed one third of that of the test tube. The Solvent B, is then added to the solution until the solution becomes cloudy. Generally, no more than twenty drops of Solvent B is needed.

Alternatively, the solid can be suspended in the second Solvent B. Then hot Solvent A is added until the solid just dissolves.

The last two steps (cooling the solution and filtering and drying the crystals) in two-solvent recrystallization are similar to those in the single-solvent method, although here to wash the crystals in the last step, use a mixture of the solvent system in about the same ratio used to obtain a saturated solution.

New words

recrystallization [rekrɪstələɪ'zeɪʃn] *n.* 再结晶作用，重结晶，重结晶作用
solubility [ˌsɒljʊ'bɪlətɪ] *n.* 溶（解）度；（可）溶性；可解释性
insoluble [ɪn'sɒljəbl] *adj.* 不能解决的，不溶的
impurity [ɪm'pjʊərətɪ] *n.* 污点，污染；掺杂，不纯；不道德，罪恶；混杂物，粗劣品
filtration [fɪl'treɪʃn] *n.* 过滤；筛选；滤清；滤除
supersaturate [suːpə'sætʃəreɪt] *vt.* 使过度饱和
hexane [hek'seɪn] *n.* （正）己烷
methanol ['meθənɒl] *n.* 甲醇
ethyl ['eθɪl] *n.* 乙荃，乙烷基；乙基

ethyl acetate ['eθɪl'æsɪˌteɪt] *n.* 乙酸乙酯
rack [ræk] *n.* 行李架；支架；刑架；（羊、猪等带前肋的）颈脊
　　　　　vt. 使痛苦，使焦虑；剥削，榨取；用刑拘折磨
clamp [klæmp] *vt. & vi.* 夹紧，夹住；锁住；把（砖等）堆高，堆存；脚步很重地走
　　　　　n. 钳，夹子；压板，压铁；车轮锁；（砖等的）堆
　　　　　vt. 紧紧抓住；紧夹住；被抓住；被夹紧
spatula ['spætʃələ] *n.* （搅拌或涂敷用的）铲，漆工抹刀；<医>压舌板
finger cot 手指套
drop-wise ['drɔpˌwaɪz] *adv.* 逐滴地，一滴一滴地
precipitate [prɪ'sɪpɪteɪt] *vt.* 下掷，由高处抛下；使提前或突然发生；[气象] 使凝结而下降；[化]使沉淀
　　　　　vi. [气象]凝结；[化]沉淀；倒落，一头栽下
　　　　　adj. 仓促行进的，匆促的；轻率的，冲动行事的水蒸气，降水；突然发生的
　　　　　n. [化]沉淀物；结果，产物
laureate ['lɒrɪət] *n.* 资金[荣誉]获得者
induce [ɪn'djuːs] *vt.* 引诱；引起；[电]感应；归纳
nucleation [ˌnjuːklɪ'eɪʃən] *n.* 成核现象，晶核形成
meniscus [mə'nɪskəs] *n.* 新月，半月板
vacuum ['vækjuəm] *n.* 真空；空白；空虚；清洁
　　　　　v. 用真空吸尘器清扫
chilled [tʃɪld] *adj.* 已冷的，冷硬了的，冷冻的
　　　　　v. （使）变冷；使很冷；使冰冷；使恐惧
adapter [ə'dæptə] *n.* 适配器；改编者；改写者；适应者；适应物
moisten ['mɔɪsn] *vt. & vi.* （使）变得潮湿，变得湿润
swirl [swɜːl] *vi.* 旋转，打旋；眩晕；盘绕；大口喝酒
　　　　　vt. 使成漩涡；使眼花；打转；弯曲盘旋
　　　　　n. （水，风等的）旋转，漩涡；<美>弯曲；卷状的东西
suction ['sʌkʃn] *n.* 吸；抽吸；吸出；相吸
　　　　　v. 抽吸；吸出
volatility [ˌvɒlə'tɪlətɪ] *n.* 挥发性；挥发度；反复无常
miscible ['mɪsəbl] *adj.* 易混合的
tare [teə(r)] *n.* 皮重；<植>巢菜；（常复数）稗子；（常复数）不良成分
　　　　　vt. 量皮重

Unit 13   Acid-Base Titration

A volumetric quantitative analytical technique that is often used to measure how much acid or base is present in a solution is called a titration. Acid-base titrations are based on neutralization reactions. If a solution is acidic, a titration is to add a base to it until the base neutralizes all the acid.

Acid-base titrations can be used for most acids and bases, including hydrochloric acid, sulfuric acid, acetic acid, sodium hydroxide, ammonia, and so on. In particular, it is even possible to determine in one titration the composition of a mixture containing acids or bases of different strengths, such as sodium hydroxide and sodium hydrogen carbonate. Hydrochloric acid and sodium hydroxide are two most commonly used reagents in acid-base titrations.

The reaction follows a stoichiometric relationship. The stoichiometric point in an acid-base titration may be visually determined by use of an indicator which tells us when the titration is completed. Visual detection of completion of the reaction is a key factor in maintaining the simplicity of titration. A visual indicator is an organic compound that changes color when the pH of the solution changes. Such pH-dependent color changes are the result of chemical changes in the indicator with its chemical environment caused by the addition of H_3O^+ or OH^-. An example of these changes in the functional moieties of phenolphthalein, a commonly used indicator which changes from colorless to a pink hue at pH 8.0-9.0.

Ideally, the observation of a sudden change in the color of the solution with the addition of a few drops of indicator tells us the completion of the titration. Sometimes color change seems like instant, with a very small drop of the titrant completely changing the color of the solution. However, depending on the concentration of titrant, titrated substance, and the selected indicator, sometimes we have to add even several milliliters of titrant before we see a color change. This confusion makes it difficult for us to determine when we should stop the titration. In some cases, we should look for a completely different indicator if the one selected fails to guarantee accuracy in the measurement. In order to choose a suitable indicator for an acid-base titration, we need to know the pH of the end point before using standard indicator tables. At the end point of the titration the pH of the solution suddenly changes. The end-point pH can be calculated with the aid of the titration equation.

As the example to show the general procedures in an acid-base titration, sodium hydroxide solution is used to titrate a solid acid dissolved in deionized water. The end point is determined by the color change of an indictor.

Designed for classical quantitative volumetric analysis, this experiment serves as a good example of conducting a quantitative experiment with the combination of several quantitative techniques. It is one of the roost accurate procedures yet one of the simplest in chemistry lab work.

In general, a solution of the acid A is added to an Erlenmeyer flask. A buret is filled with titrant, the solution of base B, at the beginning of the titration to its maximum capacity. The volume of the base solution is read before the beginning of the titration. Solution B is then added drop-wise from the buret to Solution A in the Erlenmeyer flask. The titration is completed when the indicator exhibits a permanent color change. The buret is read again to obtain the volume of Solution B added. With the known concentration of B in the titrant, the titrant volume that reacts with all of A in the flask can be used to calculate how much A is present based on stoichiometry. Some of the chemicals and equipment involved in this experiment include a buret, a buret clamp, a pipet, a small funnel, a standardized sodium hydroxide solution, and phenolphthalein indicator.

The acid-base titration starts with the dissolution of the solid acid sample in deionized water. Add the end-point indicator, which is phenolphthalein, with two drops to each flask containing the acid sample and deionized water. Properly label the flasks. Be consistent in all of the samples when adding the indicator. Swirl the flasks until the solid acid is completely dissolved. Finally, rinse with deionized water three times around, which is critical to ensure that all solid acid has been removed from the flask walls and dissolved in the solution. All solid particles must be dissolved prior to the titration.

The buret needs to be checked for if it is quantitatively clean, both to avoid contamination and to be sure that titrant volumes are accurately read. Make sure the buret stopcock is closed. Fill the buret with water and then drain it to check the buret, making sure that its walls drain cleanly. Before checking for drainage, wait a minute or two after completely draining the buret. Sometimes droplets appear on the inner walls of the buret after some time, indicating that the buret is not quantitatively clean. In this case, it is necessary to use standard cleaning procedures to clean the buret. If the buret is droplet free, then it is quantitatively clean and can be used for titration.

After the buret is cleaned, it is necessary to rinse it with the titrant, sodium hydroxide solution. Use a clean and dry funnel to add titrant to the buret. Titrant can also be poured into the buret directly with the buret removed from its holder. Small portions of titrant are used to rinse the buret in order to conserve the titrant. Hold the buret on its side and roll it to rinse the internal buret walls thorough. The buret tip is rinsed with the buret held over a waste container or sink and all the liquid being allowed to pass through the tip. Remove the last drop of titrant and continue with rinsing. Usually three times of rinsing is needed to remove any deionized water left in the buret.

Titrant can then be filled in the well-rinsed buret. Still use a funnel to add titrant to the buret. Carefully lift the funnel for smooth delivery and to avoid overfilling of the titrant. Similar to the cleaning of the buret, an alternative is to remove the buret from the buret holder, and directly pour the titrant from the titrant bottle. Let some titrant run through the buret tip into the water container and check whether there are any air bubbles in the tip. The bubbles will cause difficulty in obtaining accurate values of volume if they are not removed. The bubbles can be shaken out by opening the stopcock, firmly holding the buret with both hands, and jerking downward a bit. When bubbles are removed, tip off the hanging titrant drop and mount the buret for titration.

With the titrant filled in the buret, the samples and the buret are ready for titration. First, the initial level of the meniscus should be read. Look directly at the meniscus, and measure the meniscus at eye level from the center of the meniscus. It is critical to use a consistent buret reading procedure throughout the experiment. Use a contrast card to assist in reading the buret consistently. As a standard practice, the reading of the meniscus level should be immediately recorded in a permanent lab book. Taking notes on a scratch paper is not a correct way of recording such a critical observation.

Titration starts with sample flask number 1. Place a white paper beneath the titration flasks to aid in judging the end-point color. The buret is positioned in such a way that its tip is a few centimeters below the flask rim. The sample is titrated, using the disappearance of the indicator color as a guide of the titration rate. At the beginning of the titration, allow the titrant to run full bore into the flask. At the point where the titrant hits the acid solution the color may temporarily turn pink, but this color disappears upon swirling. The color disappearance is very rapid because of the fast production of colorless water by reaction of the base from the buret with the acid in the sample. During the titration process, continuously swirl to ensure proper mixing which leads to fast reaction.

As the rate of color changes slows, titrant can be added more slowly. With more sodium hydroxide from the buret added to the solution in the flask, more acid in the sample is reacted and less acid is available in the solution. When the red indicator color lingers in the flask for a second or two on swirling after addition of titrant, the end point is near and the delivery of titrant should be slowed down. Smaller volumes of titrant should be added carefully into the flask. The rapid addition rate at the start of the titration is consistent with the rapid indicator color change at the start of the titration, so is the slow addition near the end point and slow indicator color change near the equivalence point.

As the end point is approaching, the addition of titrant is reduced to a few drops. It requires patience and skill to locate the correct end point. Carefully watch the rate of color change. The addition should be even smaller if it takes longer for the color to fade away. Continue with ever smaller increment addition of titrant. Rinse the inner flask walls and the buret tip to make sure that no droplets remain in those places and that all the acid is in the solution. The end point is reached when the addition of a final half-drop leads to a persistent color change. The first appearance of a permanent pink coloration indicates the end point, and the solution should appear extremely pale.

Then read the final level of the meniscus in the buret and also record the reading immediately in the lab book. The difference between the initial volume and the volume left in the buret at the end of titration is the volume of the base consumed. It needs to be pointed out again that consistent reading of the buret is important. Be careful not to add too much titrant. If too much base is added and the indicator in the flask becomes deep pink or purple, an error called overtitration occurs. The entire titration needs to be repeated with a new sample.

After the first sample is done with titration, repeat the procedures for the other samples. The average value of the titrant volumes can then be used to calculate the concentration of acid in the sample with the aid of the titration equation. Note that each titration should be an independent measurement. The first or even a rough titration allows the quick determination of the approximate volume of titrant needed to neutralize the acid. Such knowledge can be used to estimate the end point for each sample. However, the predicted end point is only a guide. It should not be the target of the titration.

New words

volumetric [ˌvɒljʊ'metrɪk] *adj.* 测定体积的
titration [tɪ'treɪʃn] *n.* 滴定
acidic [ə'sɪdɪk] *adj.* （味）酸的；[化]酸的，酸性的；含有大量硅酸的
acetic [ə'sɪːtɪk] *adj.* 醋的，乙酸的
reagent [rɪ'eɪdʒənt] *n.* 反应物，试剂
deionize [diː'aɪənaɪz] *v.* 除去离子；去电离
rinse [rɪns] *vt.* 漂洗；冲洗；漂净；冲掉
    n. 冲洗；染发剂；漂洗
contamination [kənˌtæmɪ'neɪʃən] *n.* 污染；弄脏；毒害；玷污
rim [rɪm] *n.* （圆形器皿的）边，缘，框；轮缘
   vt. 环绕（圆形或环形物的）边缘；镶边
   vt. & vi. 形成……的边沿，给……镶边
linger ['lɪŋgə(r)] *vi.* 逗留，徘徊；缓慢消失
increment ['ɪŋkrəmənt] *n.* 增长；增量；增额；定期的加薪

Unit 14 Isolation, Purification, and Identification of Caffeine

The objectives of this experiment are: to isolate caffeine from tea leaves through extraction; to purify the crude material by sublimation; to identify the purified substance by measuring its melting point.

14.1 Background

Caffeine has the molecular formula $C_8H_{10}N_4O_2$ with a molecular weight of 194.19 g/mol. Its chemical name is 3,7-dihydro-1,3,7-trimethyl-lH-purine-2,6-dione. Caffeine belongs to a group of compounds called alkaloids, more specifically, a member of the methylxanthines. The caffeine molecule has base characteristics (alkali-like) and the purine ring system, which is an important framework in living systems.

Caffeine is a chemical with a variety of uses. From medicines to beverages to foods, caffeine is one of the most popular natural products used today. It is the most widely used of all the stimulants and acts to stimulate the heart, central nervous system, and the respiratory system. Its usage can increase blood pressure, contraction force, and volume output by increasing heart rate. A small dose of this compound at an amount of 50 to 200 mg increases alertness and reduces drowsiness and fatigue. Caffeine is the main ingredient of many "stay-awake" pills. It is a smooth muscle relaxant and a diuretic. Caffeine is also a food additive. It can be found in popular soft drinks. However, it needs to be pointed out that caffeine has side effects. Large doses in excess of 200 mg can cause insomnia, restlessness, headaches, and muscle tremors. In addition, continued, heavy use of this chemical may lead to addictiveness. Furthermore, some research connects high caffeine consumption in pregnant women with the malformation of their children.

As a natural product, caffeine constitutes as much as 5% by weight of tea and coffee leaves, and is also present in cola nuts and cacao beans. It can be isolated from these natural sources through a process known as extraction, which is a chemical method of separating a specific component of a solution from the rest of the solution. This method is done by taking advantage of the solubility characteristics of a particular chemical with a given solvent. Caffeine is easily soluble in organic solvents such as chloroform or dichloromethane (CH_2Cl_2), but only partially soluble in water.

In this experiment, caffeine is first separated from tea leaves using hot water since tea leaves consist primarily of cellulose which is insoluble in water. Hot water swells the tea leaves to release caffeine and other water soluble compounds such as tannins, complex substances which are colored phenolic compounds of high molecular weight. Since tannins are acidic, they can react with a basic salt such as Na_2CO_3 to form salts. These salts are soluble in water, but insoluble in organic solvents. Although caffeine is somewhat soluble in water, it is more soluble in the organic solvent dichloromethane. Therefore, dichloromethane can be used to selectively extract caffeine from its water solution. The sodium salts of the tannins remain behind in the aqueous solution. The dichloromethane solution is then passed through some solid Na_2SO_4 to remove trace water. Evaporation of dichloromethane yields crude material of caffeine, which can be further purified by sublimation.

Sublimation is a phase transformation process in which a solid converts to a gas directly without going through the liquid state. Relatively few solids possess this kind of behavior at atmospheric pressure. Caffeine is one of the few examples. Other examples are solid compounds naphthalene (mothballs), iodine, and solid carbon dioxide (dry ice).

Chemicals possess characteristic physical properties which facilitate their identification. In many cases, a thorough determination of the physical properties of a given substance can be used for its identification. The physical properties of an unknown compound can be compared to properties of known substances that are tabulated in the chemical literature, and identification can be assumed if a match can be made. The physical properties most commonly listed in handbooks of chemical data include color, density, solubility in various solvents, melting point, sublimation characteristics and so on. The melting point of a compound refers to the temperature at which the solid and liquid states are in equilibrium. A pure substance usually has a quite sharp melting transition and a very narrow range of melting point. Impurities lower the melting point and cause a broadening of the range. The criteria for purity of a solid are the correspondence to the value in the literature and the narrowness of the melting-point range. Thus the purity of caffeine alter sublimation can be verified by its melting point. Pure caffeine forms white, hexagonal crystals, which can be ground into soft powder. Its melting point in chemical handbook is 238℃.

14.2 Procedures

Commercial tea bags are used as samples for the extraction of caffeine. The experiment follows the scheme shown before.

14.2.1 Isolation of caffeine by extraction

Without tearing the paper, open two tea bags with care. Weigh the contents to the nearest 1 mg and record this weight. Put the tea leaves back into the bags. Then close and securely seal the bags with staples. Place the tea bags in a 150-mL beaker and let the bags lie flat at the bottom. Add 30 mL of distilled water and 2.0 g of anhydrous Na_2CO_3 into the beaker. Heat the water on a hot plate to a gentle boiling. Cover the beaker with a watch glass and continue heating for approximately 15 minutes. Keep the tea bags under water by occasionally pushing them back down with a glass rod, making sure that the tea leaves are covered with as much hot water as possible. Watch for loss of water, additional water may be needed.

Pour the hot, concentrated tea extract into a 50 mL Erlenmeyer flask. Add 10 mL of hot water and carefully press the tea bags with a glass rod. Be careful not to break the tea bags, since the presence of tea leaves in the solution will lead to additional difficulties in the separation process. Add the wash water to the tea extract in the flask. Filter any solids present in the tea extract by gravity. Discard the tea bags. Cool the combined tea solution using an ice-water bath.

Transfer the cool tea extract from the flask to a 125 mL separatory funnel supported on a ring stand with a ring clamp. Into the funnel, add 5.0 mL of dichloromethane. Stopper the funnel and lift it with two hands. Hold the stopper in place with one hand and invert the funnel to gently mix the contents three to four times. Be sure that the liquid is not in contact with the stopcock. When the funnel is inverted for mixing, open the stopcock to release any pressure built up by the volatile solvent. Always point the opening away from any person.

Put the separatory funnel back to the ring clamp. Remove the stopper and let the aqueous layer settle and separate from the dichloromethane layer, resulting in two distinct layers after a few minutes. Carefully manipulate the stopcock and drain the dichloromethane layer at the bottom into a 25 mL Erlenmeyer flask. Try not to transfer any of the aqueous solution along with the organic layer. Add a fresh 5.0 mL of dichloromethane and repeat the extraction. Combine the separated bottom dichloromethane layers. Dry the combined extract by adding 0.5 g of anhydrous Na_2SO_4. Swirl the flask for better performance.

Using a gravity filtration, filter the dichloromethane — salt mixture to a clean pre-weighed 25 mL side-arm filter flask containing one or two boiling stones. Rinse the salt on the filter paper with an additional 2.0 mL of dichloromethane: Gently heat the flask in a hot water bath to remove the dichloromethane by evaporation. The solid residue is crude caffeine. Weigh the flask to get its weight and determine the percentage yield.

14.2.2 Purification of caffeine with sublimation

The caffeine is now purified by sublimation, which is done with a cold finger condenser as the sublimation setup. Using some glycerin as a lubricant, carefully insert the cold finger condenser into a suitable neoprene adapter. Adjust the position of the cold finger so that its tip is 1 cm from the bottom of the side-arm filter flask containing the crude caffeine. Clean any remaining glycerin on the cold finger and dry the cold finger surface.

Connect the cold finger to a faucet with latex tubing and also connect the side-arm filter flask to a vacuum pump by vacuum tubing. Install a trap between the pump and the sublimation setup. When turning on the water, press the cold finger into the filter flask to ensure a good seal. Heat the sample gently and carefully with a microburner to sublime the caffeine. Hold the base of the microburner and move the flame around the flask. Avoid melting the sample. If the caffeine happens to melt, stop heating and allow it to cool before continuing the sublimation. When the sublimation is complete, discontinue heating and allow the system to cool while the vacuum is still on. After the system has cooled, remove the vacuum and carefully collect the purified caffeine from the cold finger.

Weigh the purified solid and calculate the percentage of caffeine in the tea sample.

14.2.3　Identification of caffeine by its melting point

Determine the melting point of the purified solid using a Thiele tube. First, collect a sample of the caffeine in a capillary tube and seal the tube with a torch afterwards. Attach the sealed melting-point capillary tube to the thermometer using a rubber ring, making certain that the tip of the melting-point capillary containing the solid is next to the mercury bulb of the thermometer. Support the Thiele tube on a ring stand with an extension clamp. Into the Thiele tube, fill silicone oil to a level which is above the top of the side arm. Using a thermometer clamp, support the thermometer with the attached melting-point capillary tube in the oil. Immerse the bulb and capillary tube in the oil, but keep the rubber ring and top of the capillary tube out of the oil.

Very slowly, heat the arm of the Thiele tube with a burner by using a small flame and gently moving the burner along the arm of the Thiele tube. Record the temperature when the solid begins to liquefy and the temperature when the solid is completely a liquid. These two temperatures define the melting-point range.

Compare the melting point obtained with the literature value.

New words

sublimation [ˌsʌblɪ'meɪʃn] *n.* 升华，升华物，高尚化
alkaloids ['ælkəlɔɪdz] *n.* [医]半边莲属碱，类生物碱类，生物碱类
methylxanthine ['meθɪlksænθaɪn] *n.* 甲基化黄嘌呤衍生物
purine ['pjʊrɪn] *n.* 嘌呤，咖啡碱
stimulant ['stɪmjələnt] *n.* [药]兴奋剂；刺激物；酒精饮料
　　　　　　　　　　adj. 刺激的，激励的；使人兴奋的
contraction [kən'trækʃn] *n.* 收缩，缩减
drowsiness ['draʊzɪnəs] *n.* 睡意；嗜睡
diuretic [ˌdaɪju'retɪk] *n.* [医]利尿剂
　　　　　　　　　　adj. 利尿的
additive ['ædətɪv] *n.* 添加剂；添加物；[数]加法
　　　　　　　　　　adj. 附加的；[化]加成的；[数]加法的
insomnia [ɪn'sɒmnɪə] *n.* [医]失眠，失眠症
restlessness ['restləsnəs] *n.* 坐立不安，心神不定，无休止

tremor ['tremə] *n.* 震颤；战栗；震颤声；大地的轻微震动
malformation [ˌmælfɔː'meɪʃn] *n.* 难看，畸形
cola ['kəʊlə] *n.* 可乐果树；可乐饮料
cacao [kə'kaʊ] *n.* 可可，可可树，可可豆
chloroform ['klɒrəfɔːm] *n.* （用作麻醉剂的）氯仿，三氯甲烷
　　　　　　vt. 用氯仿麻醉
dichloromethane [daɪklɔːrə'meθeɪn] *n.* 二氯甲烷
cellulose ['seljuləʊs] *n.* 细胞膜质，纤维素；（用于制作涂料、漆等的）纤维素化合物
tannin ['tænɪn] *n.* 鞣酸，丹宁酸；鞣料；鞣质
phenolic [fɪ'nɒlɪk] *adj.* 酚的，石碳酸的
naphthalene ['næfθəliːn] *n.* 萘（球），卫生球
mothball ['mɒθbɔːl] *n.* 卫生球，樟脑球
　　　　　　vt. 封存
　　　　　　adj. 后备的
iodine ['aɪədiːn] *n.* <化>碘
tabulate ['æbjuleɪt] *v.* 把（数字、事实）列成表；使成板[片]状，使成平面
　　　　　　adj. 平板状的，有平面的，[动物，动物学]有横隔板的
literature ['lɪtrətʃə(r)] *n.* 文学；文学作品；文献；著作
equilibrium [ˌiːkwɪ'lɪbrɪəm] *n.* 平衡，均势；平静
verify ['verɪfaɪ] *vt.* 核实；证明；判定
hexagonal [heks'ægənl] *adj.* 六角形的，六边形的；六方
grind [graɪnd] *vt. & vi.* 磨碎，嚼碎；折磨；（过去式和过去分词：ground）
　　　　　　vt. 磨快，磨光；咬牙
　　　　　　vi. 嘎吱嘎吱地擦
　　　　　　n. 碾，磨；苦差事，苦活儿；〈美〉刻苦用功的学生；研细的程度
staple ['steɪpl] *n.* 主要产品；钉书钉，U形钉；主题，主要部分；主食
　　　　　　vt. 用钉书钉钉住
　　　　　　adj. 最基本的，最重要的
distill [dɪs'tɪl] *v.* 蒸馏，提取，滴下
　　　　　　vt. 抽出……的精华；提炼；使渗出；使滴下
　　　　　　vi. 蒸馏；精炼；作为精华产生；渗出
anhydrous [æn'haɪdrəs] *adj.* 无水的（尤指结晶水）
gravity ['grævətɪ] *n.* 重力；万有引力，地心引力；重要性，严重性；严肃，庄重
discard [dɪs'kɑːd] *vt.* 丢弃，抛弃；解雇；出牌
　　　　　　n. 被抛弃的人[物]；丢弃，抛弃；打出的牌；打出的牌
　　　　　　vi. 出无用的牌；垫牌
stopper ['stɒpə] *n.* 阻塞物，（尤指）瓶塞
　　　　　　vt. （用瓶塞）塞住

volatile ['vɒlətaɪl] *adj.* 易变的，不稳定的；易挥发的；爆炸性的；快活的，轻快的
finger condenser ['fɪŋgə kən'densə] *n.* [医]指形冷凝管
glycerin ['glɪsərɪn] *n.* <化>甘油，丙三醇
lubricant ['lu:brɪkənt] *n.* 润滑剂，润滑油；能减少摩擦的东西
　　　　　　　　　　adj. 润滑的
neoprene ['nɪːəpriːn] *n.* 氯丁（二烯）橡胶
faucet ['fɔːsɪt] *n.* <美>水龙头
latex ['leɪteks] *n.* 胶乳，（尤指橡胶树的）橡浆；人工合成胶乳
trap [træp] *vt.* 诱骗；使受限制；困住；使（水与气体等）分离
　　　　　　n. 圈套；（对付人的）计谋；（练习射击用的）抛靶器；（捕捉动物的）夹子
　　　　　　vi. 设陷阱；装捕捉机；设圈套
microburner ['maɪkrəʊbɜːnər] *n.* 微型燃烧器
capillary [kə'pɪlərɪ] *n.* 毛细管；毛细血管；微管
　　　　　　　adj. 毛细管的；毛状的；表面张力的；细长的
mercury ['mɜːkjərɪ] *n.* [化]汞，水银；[天]水星；温度表；精神，元气

Part 4

Academic Reading

Unit 15 A (R) evolution in Chemistry

The power of evolution is revealed through the diversity of life. The Nobel Prize in Chemistry 2018 is awarded to Frances H. Arnold, George P. Smith and Sir Gregory P. Winter for the way they have taken control of evolution and used it for the greatest benefit to humankind. Enzymes developed through directed evolution are now used to produce biofuels and pharmaceuticals, among other things. Antibodies evolved using a method called phage display can combat autoimmune diseases and, in some cases, cure metastatic cancer.

We live on a planet where a powerful force has become established: evolution. Since the first seeds of life appeared around 3.7 billion years ago, almost every crevice on Earth has been filled by organisms adapted to their environment: lichens that can live on bare mountainsides, archaea that thrive in hot springs, scaly reptiles equipped for dry deserts and jelly fish that glow in the dark of the deep oceans.

In school, we learn about these organisms in biology, but let's change perspective and put on a chemist's glasses. Life on Earth exists because evolution has solved numerous complex chemical problems. All organisms are able to extract materials and energy from their own environmental niche and use them to build the unique chemical creation that they comprise. Fish can swim in the polar oceans thanks to antifreeze proteins in their blood and mussels can stick to rocks because they have developed an underwater molecular glue, to give just a few of the innumerable examples.

The brilliance of life's chemistry is that it is programmed into our genes, allowing it to be inherited and developed. Small random changes in genes change this chemistry. Sometimes this leads to a weaker organism, sometimes a more robust one. New chemistry has gradually developed and life on Earth has become increasingly complex.

Fig. 15-1 The Nobel Laureates in Chemistry 2018 have harnessed evolution and further advanced it in their laboratories.

This process has now come so far that it has given rise to three individuals so complex they have managed to master evolution themselves. The Nobel Prize in Chemistry is awarded to Frances H. Arnold, George P. Smith and Sir Gregory P. Winter, because they have revolutionised both chemistry and the development of new pharmaceuticals through directed evolution. Let's begin with the star of enzyme engineering: Frances Arnold.

15.1 Enzymes—the sharpest chemical tools of life

Even in 1979, as a newly graduated mechanical and aerospace engineer, Frances Arnold had a clear vision: to benefit humanity through the development of new technology. The US had decided that 20 per cent of its power would come from renewable sources by 2000, and she worked with solar power. However, prospects for the future of this industry changed radically after the 1981 presidential election, so instead she turned her gaze to the new DNA technology. As she expressed it herself: "It was clear that a whole new way of making materials and chemicals that we needed in our daily lives, would be enabled by the ability to rewrite the code of life."

Instead of producing pharmaceuticals, plastics and other chemicals using traditional chemistry, which often requires strong solvents, heavy metals and corrosive acids, her idea was to use the chemical tools of life: enzymes. They catalyse the chemical reactions that occur in the Earth's organisms and, if she learned to design new enzymes, she could fundamentally change chemistry.

15.2 Human thought has limitations

Initially, like many other researchers at the end of the 1980s, Frances Arnold attempted to use a rational approach to rebuilding enzymes to give them new properties, but enzymes are extremely complex molecules. They are built from 20 different kinds of building blocks—amino acids—that can be infinitely combined. A single enzyme can consist of several thousand amino acids. These are linked together in long chains, which fold up into special three-dimensional structures. The environment necessary to catalyse a particular chemical reaction is created inside these structures.

Using logic to try to work out how this elaborate architecture should be remodelled to give an enzyme new properties is difficult, even with contemporary knowledge and computer power. In the early 1990s, humble in the face of nature's superiority, Frances Arnold decided to abandon this, in her words, "somewhat arrogant approach" and instead found inspiration in nature's own method for optimising chemistry: evolution.

15.3 Arnold starts to play with evolution

For several years, she had tried to change an enzyme called subtilisin so that rather than catalyzing chemical reactions in a water-based solution, it would work in an organic solvent, dimethylformamide (DMF). Now she created random changes—mutations—in the enzyme's genetic code and then introduced these mutated genes into bacteria that produced thousands of different variants of subtilisin.

After this, the challenge was to find out which of all these variants worked best in the organic solvent. In evolution, we talk about survival of the fittest; in directed evolution this stage is called selection.

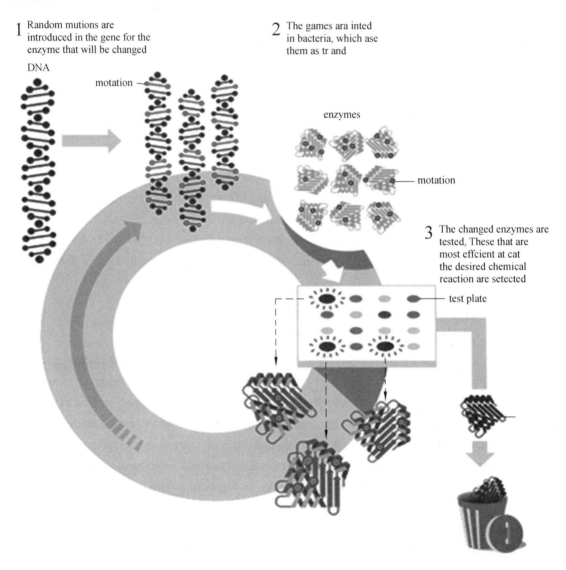

Fig. 15-2 The underlying principle for the directed evolution of enzymes. After a few cycles of directed evolution, an enzyme may be several thousand times more effective.

Frances Arnold utilised the fact that subtilisin breaks down milk protein, casein. She then selected the variant of subtilisin that was most effective in breaking down casein in a solution with 35 percent DMF. She subsequently introduced a new round of random mutations in this subtilisin, which yielded a variant that worked even better in DMF.

In the third generation of subtilisin she found a variant that worked 256 times better in DMF than the original enzyme. This variant of the enzyme had a combination of ten different mutations, the benefits of which no one could have worked out in advance.

With this, Frances Arnold demonstrated the power of allowing chance and directed selection, instead of solely human rationality, to govern the development of new enzymes. This was the first and most decisive step towards the revolution we are now witnessing.

The next important step was taken by Willem P. C. Stemmer, a Dutch researcher and entrepreneur who died in 2013. He introduced yet another dimension to the directed evolution of enzymes: mating in a test tube.

15.4 Mating—for more stable evolution

One prerequisite for natural evolution is that genes from different individuals are mixed through mating or pollination, for example. Beneficial properties can then be combined and give rise to a more robust organism. At the same time, less functional gene mutations can disappear from one generation to another.

Willem Stemmer used the test tube equivalent to mating: DNA shuffling. In 1994 he demonstrated that it was possible to cut different versions of a gene into small pieces and then, helped by the tools of DNA technology, puzzle together the pieces into a complete gene, one that is a mosaic of the original versions.

Using several cycles of DNA shuffling, Willem Stemmer changed an enzyme so that it became much more effective than the original enzyme. This showed that mating genes together—researchers call this recombination—can result in the even more efficient evolution of enzymes.

15.5 New enzymes produce sustainable biofuel

The tools of DNA technology have been refined since the early 1990s, and the methods used in directed evolution have multiplied. Frances Arnold has been at the leading edge of these developments; the enzymes now produced in her laboratory can catalyse chemistry that does not even exist in nature, producing entirely new materials. Her tailored enzymes have also become important tools in the manufacture of various substances, such as pharmaceuticals. Chemical reactions are sped up, produce fewer by-products and, in some cases, it has been possible to exclude the heavy metals required by traditional chemistry, thus considerably reducing environmental impact.

Things have also come full circle: Frances Arnold is again working with the production of renewable energy. Her research group has developed enzymes that transform simple sugars to isobutanol, an energy-rich substance that can be used for the production of both biofuels and greener plastics. One long-term aim is to produce fuels for a more environmentally friendly transport sector. Alternative fuels—produced by Arnold's proteins—can be used in cars and aeroplanes. In this way, her enzymes are contributing to a greener world.

And now to the second half of 2018's Nobel Prize in Chemistry, where directed evolution has instead led to pharmaceuticals that can neutralise toxins, combat the progression of autoimmune diseases and, in some cases, even cure metastatic cancer. This is where a vital role is played by a tiny virus that infects bacteria and the method known as phage display.

15.6 Smith uses bacteriophages

As is so often the case, science took an unpredictable path. In the first half of the 1980s, when George Smith started to use bacteriophages—viruses that infect bacteria—it was mainly in the hope that they could be used to clone genes. DNA technology was still young and the human genome was like an undiscovered continent. Researchers knew that it contained all the genes required to produce the body's proteins, but identifying the specific gene for a particular protein was more difficult than looking for a needle in a haystack.

However, there were huge benefits for the scientist who did find it. Using that time's new genetic tools, the gene could be inserted into bacteria which—with a bit of luck—could mass produce the protein to be studied. The whole process was called gene cloning and George Smith's idea was that researchers looking for genes could use bacteriophages in an ingenious way.

15.7 Bacteriophages—a link between a protein and its unknown gene

Bacteriophages are simple by nature. They consist of a small piece of genetic material that is encapsulated in protective proteins. When they reproduce, they inject their genetic material into bacteria and hijack their metabolism. The bacteria then produce new copies of the phage's genetic material and the proteins that form the capsule, which form new phages.

George Smith's idea was that researchers should be able to use the phages' simple construction to find an unknown gene for a known protein. At this time, large molecular libraries were available, which contained masses of fragments of various unknown genes. His idea was that these unknown gene fragments could be put together with the gene for one of the proteins in the phage capsule. When new phages were produced, the proteins from the unknown gene would end up on the surface of the phage as part of the capsule protein (figure 15-3).

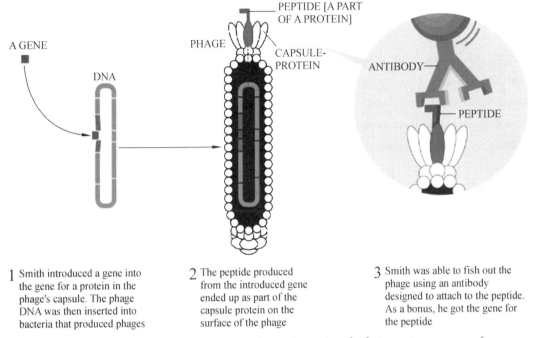

1. Smith introduced a gene into the gene for a protein in the phage's capsule. The phage DNA was then inserted into bacteria that produced phages

2. The peptide produced from the introduced gene ended up as part of the capsule protein on the surface of the phage

3. Smith was able to fish out the phage using an antibody designed to attach to the peptide. As a bonus, he got the gene for the peptide

Fig. 15-3 Phage display — George Smith developed this method for finding unknowns genes for a known protein.

15.8 Antibodies can fish out the right protein

This would result in a mixture of phages that carried multitudes of different proteins on their surface. In the next stage—George Smith postulated—researchers would be able to use antibodies to fish phages carrying various known proteins out of this soup. Antibodies are proteins that function like targeted missiles; they can identify and bind to a single protein among tens of thousands of others with extreme precision. If researchers caught something in the phage soup using an antibody that they knew attached to a known protein, as a bycatch they would get the thus-far unknown gene for the protein.

It was an elegant idea and, in 1985, George Smith demonstrated that it could work. He produced a phage that carried part of a protein, a peptide, on its surface. Using an antibody, he was then able to fish the phage he had constructed out of a soup of many phages.

Through this experiment, George Smith laid the foundation of what has come to be known as phage display. The method is brilliant in its simplicity. Its strength is that the phage functions as a link between a protein and its gene. However, it was not within gene cloning that the method had its major breakthrough; instead, in around 1990, several research groups started to use phage display to develop new biomolecules. One of the people who adopted the technique was Gregory (Greg) Winter and it is thanks to his research that phage display is now bringing great benefit to mankind. To understand why, we need to take a closer look at antibodies.

15.9 Antibodies can block disease processes

The human lymphatic system has cells that can produce hundreds of thousands of different kinds of antibodies. Using a well-developed system, all these cells have been tested so that no antibody attaches to any of the body's various types of molecules. However, this enormous variation ensures that there is always an antibody that attaches to the viruses or bacteria that infect us. When an antibody attaches to them it sends a signal to aggressive immune cells to destroy the invaders.

Because antibodies are highly selective and can attach themselves to a single molecule among tens of thousands of others, researchers had long hoped it would be possible to design antibodies that block various disease processes in the body and function like pharmaceuticals. Initially, to obtain these therapeutic antibodies, mice were injected with different targets for pharmaceuticals, such as proteins from cancer cells. However, in the 1980s it became increasingly clear that this method had limitations; some substances were toxic for the mice and others did not result in any antibody production. Additionally, it was discovered that the obtained antibodies were recognised as foreign by the patients' immune system, which attacked them. This led to the mice antibodies being destroyed and there was a risk of side effects for the patients.

It was this obstacle that caused Greg Winter to start investigating the potential of George Smith's phage display. He wanted to avoid using mice and to be able to base pharmaceuticals on human antibodies because they are tolerated by our immune system.

15.10 Winter puts antibodies on the surface of phages

Antibodies are Y-shaped molecules; it is the far end of each arm that attaches to foreign substances. Greg Winter joined the genetic information for this part of the antibody to the gene for one of the phage's capsule proteins and, in 1990, he demonstrated that this led to the antibody's binding site ending up on the surface of the phage. The antibody he used was designed to attach to a small molecule known as phOx. When Greg Winter used phOx as a kind of molecular fishing hook, he succeeded in pulling the phage with the antibody on its surface out of a soup of four million other phages.

After this, Greg Winter showed that he could use phage display in the directed evolution of antibodies. He built up a library of phages with billions of varieties of antibodies on their surfaces. From this library, he fished out antibodies that attached to different target proteins. He then randomly changed this first generation of antibodies and created a new library, in which he found antibodies with even stronger attachments to the target. For instance, in 1994 he used this method to develop antibodies that attached to cancer cells with a high level of specificity.

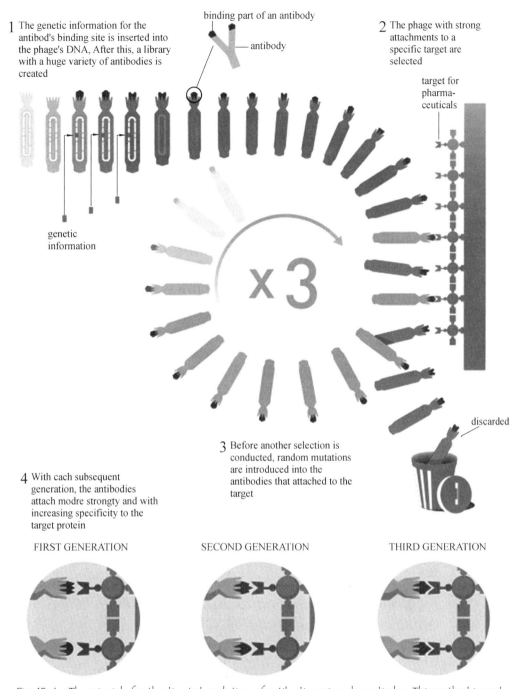

Fig. 15-4 The principle for the directed evolution of antibodies using phage display. This method is used to produce new pharmaceuticals.

15.11 The world's first pharmaceutical based on a human antibody

Greg Winter and his colleagues founded a company based on the phage display of antibodies.

In the 1990s, it developed a pharmaceutical entirely based on a human antibody: adalimumab. The antibody neutralises a protein, TNF-alpha, that drives inflammation in many autoimmune diseases. In 2002, the pharmaceutical was approved for the treatment of rheumatoid arthritis and is now also used for treating different types of psoriasis and inflammatory bowel diseases.

The success of adalimumab has spurred significant development in the pharmaceutical industry and phage display has been used to produce cancer antibodies, amongst others. One of these releases the body's killer cells so they can attack tumour cells. Tumour growth slows down and, in some cases, even patients with metastatic cancer are cured, which is a historic breakthrough in cancer care. Another antibody pharmaceutical that has been approved neutralises the bacterial toxin that causes anthrax, while another slows the autoimmune disease known as lupus; many more antibodies are currently undergoing clinical trials, for example to combat Alzheimer's disease.

15.12 The start of a new era in chemistry

The methods that the 2018 Nobel Laureates in Chemistry have developed are now being internationally developed to promote a greener chemicals industry, produce new materials, manufacture sustainable biofuels, mitigate disease and save lives. The directed evolution of enzymes and the phage display of antibodies have allowed Frances Arnold, George Smith and Greg Winter to bring the greatest benefit to humankind and to lay the foundation for a revolution in chemistry.

Unit 16   A Storied Russian Lab is Trying to Push the Periodic Table Past Its Limits—and Uncover Exotic New Elements

Fig. 16-1   Yuri Oganessian has contributed so much to the periodic table that element 118 is named after him. Now, he wants to find even heavier elements.

Dubna, Russia—From certain angles, the Flerov Laboratory of Nuclear Reactions here looks more like an auto repair shop than a legendary scientific institute. Scientists in dirty blue smocks walk around while an oil pump thumps out a techno beat. Tables are strewn with bolts and cleaning fluids, including a vodka bottle half full of ethanol. And spare parts are everywhere—bins, shelves, whole walls full of metal whatsits in all manner of disrepair.

All that stuff serves the lab's six particle accelerators, some of which resemble huge mechanical caterpillars, with dozens of tractor-green segments winding through entire rooms. Or multiple rooms: When equipment doesn't fit, researchers knock holes in walls and thread things through the concrete. Seeing the whole of an accelerator requires some serious gymnastika, scaling perilously steep stairs and dodging anacondas of hanging wires. The pipes you duck under bear warning signs to watch out—not for your head, but for the equipment. At Flerov, particles have the right of way.

Deservedly so. In various iterations, these accelerators have produced nine new elements on the periodic table over the past half-century, including the five heaviest known elements, up to number 118.

The man leading that work is physicist Yuri Oganessian, who has been at Flerov since Nikita

Khrushchev signed orders in 1956 to establish a secret nuclear lab in the birch forests here, 2 hours north of Moscow. Oganessian, 85, is a short man with bushy white hair whose voice squeaks when he gets excited. He wanted to study architecture in college until a bureaucratic snafu diverted him into physics. He still misses his first love: "I really need something visual with my science. I feel this deficit."

Fittingly, no living person has shaped the architecture of the periodic table more than he has, which is why element 118 is called oganesson. And he's not done yet. To push the table further, the lab has built a new $60 million facility, dubbed the Superheavy Element Factory (SHEF), which will start to hunt for element 119, 120, or both, this spring.

Some scientists argue that finding new elements is not worth the money, especially when those atoms are inherently unstable and will disappear in a blink. "I personally don't find it exciting, as a scientist, just to produce more short-lived elements," says Witold Nazarewicz, a physicist who studies nuclear structure at Michigan State University in East Lansing.

But to element hunters, the payoff is compelling. The new elements would extend the table—now seven rows deep—to an eighth row, where some theories predict exotic traits will emerge. Elements in that row might even destroy the table's very periodicity because chemical and physical properties might not repeat at regular intervals anymore. Pushing further into the eighth row also could answer questions that scientists have wrestled with since Dmitri Mendeleev's day: How many elements exist? And how far does the table go?

The decision to build the SHEF was tough in some ways. Besides the high cost, constructing the "factory" meant abandoning the old accelerators—which produced so many new elements—to other projects. "Emotionally," Oganessian says, "it's not easy to take something (offline) that gave you a lot. But there is no other way forward."

The heaviest element found in any appreciable amount in nature is uranium, atomic number 92. (The atomic number refers to the number of protons in an atom's nucleus.) Beyond that, scientists must create new elements in accelerators, usually by smashing a beam of light atoms into a target of heavy atoms. Every so often, the nuclei of the light and heavy atoms collide and fuse, and a new element is born. Slamming neon (element 10) into uranium, for example, yields nobelium (102).

But the odds of fusion (and survival) decrease markedly as atoms grow heavier because of increased repulsion between the positively charged nuclei, among other factors. Creating most elements in the superheavy realm (beyond 104) therefore requires special tricks. Oganessian developed one in the 1970s: cold fusion. Unrelated to the notorious nuclear power work of the 1980s, Oganessian's cold fusion involves uniting beam and target atoms that are more similar in size than those in traditional element making. And rather than smashing them together, "We bring two nuclei together so that it is a 'soft touching,'" Oganessian says. Doing that is harder than it sounds because the beam and target nuclei are both positively charged and therefore repel each other. Incoming atoms need enough speed to overcome that repulsion, but not so much that they blow the resulting superheavy nucleus apart.

1. Element maker

The $60 million Superheavy Element Factory (SHEF) in Dubna, Russia, aims to create new elements that extend the periodic table by colliding a beam of nuclei with a target. Compared with previous accelerators, the SHEF has a more intense beam, which is accelerated to roughly one-tenth the speed of light in a tight space.

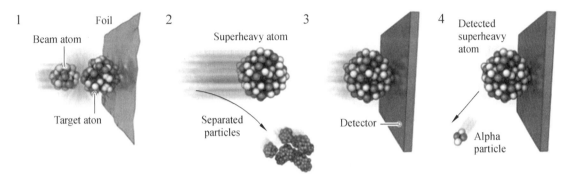

Fig. 16-2 Element maker.

A forced marriage

1. After acceleration, a beam atom crashes into a target atom sitting on a thin foil (not to scale). 2. If fusion occurs, the resulting superheavy atom flies through the foil and is separated from extraneous beam atoms. 3. The atom lands on a detector. 4. The atom sheds alpha particles, which this detector or a secondary one can sense, allowing scientists to reconstruct the atom's identity.

A team at the GSI Helmholtz Centre for Heavy Ion Research in Darmstadt, Germany, perfected Oganessian's technique and used it to create elements 107 through 112. But the method ran into limitations as the odds of fusion and survival dropped precipitously. Starting in 2003, a team at the RIKEN Institute in Wako, Japan, tried to use cold fusion to create element 113, firing zinc (element 30) onto bismuth (83). They got one atom the next year and another in 2005, which they celebrated in their control room with cheers, beers, and sake.

Then, the agony started. Needing one more atom to confirm the discovery, the RIKEN team reran the experiment in 2006 and 2007. None appeared. They tried again in 2008 and 2009. Nothing. Not until 2012—7 years later—did they detect another. "Honestly, we felt that we would be not lucky," remembers RIKEN nuclear chemist Hiromitsu Haba. "Only God knows the statistics." None of the atoms survived longer than 5 milliseconds before decaying.

Getting beyond 113 required a different approach, hot fusion, which Flerov scientists had developed in the late 1990s. Hot fusion uses higher beam energies and relies on a special isotope with a large excess of neutrons, calcium-48. (Neutrons stabilize a superheavy atom by diluting the repulsive force of protons, which would otherwise tear the nucleus apart.) Calcium-48 is expensive—it must be laboriously isolated from natural calcium sources—at $250,000 per gram. But the investment paid off. RIKEN sweated for 9 years to find three atoms of 113. Dubna snagged that many atoms of 114 within 6 months, a discovery Oganessian and colleagues celebrated in their control room with cheers, beers, and shots of spirits.

At that point, producing the next few superheavies was largely arithmetic. Calcium is element 20, and calcium plus americium (element 95) yielded element 115. Calcium plus curium (96) yielded element 116, and so on. By 2010, Dubna—in collaboration with scientists at Lawrence Livermore National Laboratory in California and Oak Ridge National Laboratory in Tennessee—had filled the periodic table's seventh row.

After 118, however, things stalled again. Fusion requires several milligrams of the target element, and producing enough einsteinium (element 99) to make element 119 is impossible with today's technology. Some researchers proposed replacing calcium-48 with titanium-50, which has two more protons, and then firing it at elements 97 and 98 to produce 119 and 120, respectively. But for technical reasons, the likelihood of fusion is just one-twentieth as high with titanium as with calcium. For most accelerators, that drops the odds of success into the realm of RIKEN's experiments to create 113—God's statistics all over again.

The SHEF was built to overcome those obstacles. In contrast to the grease monkey feel of the older Flerov accelerators, the SHEF is pristine: Bubble wrap still covers the door handles, and for now the floors are spotless.

Fig. 16-3　An office at the Flerov Laboratory of Nuclear Reactions in Dubna, Russia, preserves decades-old instruments—and an outdated periodic table.

Overall, the SHEF is a fusion of the brawny and the delicate. The beam originates in an ion source and accelerator that stands two stories high, bigger than some dachas in town. The ion source fires off 6 trillion atoms per second, 10 to 20 times as many as other element making accelerators. After twisting through a few 90° turns—the most compact arrangement in a tight space—the beam plummets into a massive cyclotron, whose very presence here is remarkable. The cyclotron's 1000-ton magnet was fabricated in 2014 in Kramatorsk, Ukraine, near the front line of the recent war with Russia, says Alexander Karpov, a Flerov physicist. The city endured heavy shelling and other military action then, and Karpov says lab personnel were nervous that the magnet would be damaged or destroyed.

After accelerating the beam to roughly one-tenth the speed of light, the cyclotron directs it toward the delicate part of the operation: micrometer-thin metallic foils with target atoms plated onto them. Those foils are mounted onto a disk roughly the size of a CD, which spins to keep cool. If it didn't, the beam would fry a hole in it.

If fusion occurs, the resulting superheavy atom sails through the foil. Unfortunately, the foil is so thin that gobs of other particles slip through as well, producing a blizzard of extraneous noise. That's when the separator comes into play. It consists of five magnets painted the same bright red as a fire truck and collectively weighing twice as much as one—64 tons. Despite the bulk, the magnets are aligned to within 0.01 millimeters, and their fields are precise enough to filter out lighter atoms, including nearly all beam atoms, swerving them into a device called the beam dump.

The separator, like the beam source, gives the SHEF an advantage. Earlier separators were tuned to superheavy atoms with a narrow range of speed, charge, and direction; those that deviated too much ended up in the beam dump. The new separator is more generous, giving a pass to two to three times as many superheavy atoms.

After slaloming through the separator, an atom arrives at a silicon-germanium detector, which records the atom's position and time of arrival and then starts to monitor it. Superheavy atoms decay by emitting a series of alpha particles—bundles of two protons and two neutrons. Releasing an alpha changes the atom's identity: element 118 becomes 116, which becomes 114, and so on.

That decay chain is what allows scientists to identify, retroactively, which element they've created. Each alpha particle in the chain flies off with a characteristic energy. So if the detector spots an alpha with the right energy—and, crucially, sees that it emerged from the same point on the detector where a superheavy atom just landed—it begins to watch closely for more alphas.

To aid that search, the detector automatically shuts off the cyclotron beam to reduce the amount of cruft flying around. The shutdown also triggers a loud beep in the SHEF's control room, where a few probably bored scientists will be sitting. (On a recent visit to another control room here, two graduate students were watching a schlocky sci-fi monster flick.) The bell is a moment of excitement amid the monotony.

It's also superfluous. Inside the detector, the atom will continue to shed alphas: In fact, several events in the decay chain will already have occurred before the scientists even register the sound. With superheavies, it's hard come, easy go. Only later—when the scientists comb through the raw data and match every detected alpha particle to a specific element in the decay chain—can they reconstruct which element they initially created.

The stronger beam and more generous separator should, in theory, cancel out the lower odds of titanium-50 fusion. That gives the Dubna team hope that atoms of 119 or 120 will soon reveal themselves. A team at RIKEN is also searching for 119, albeit using a different and perhaps harder method (firing vanadium, element 23, onto curium). Between the two labs, scientists are confident that 119 and 120 will appear somewhere within about 5 years.

It's the next 5 years that worry people. Creating elements heavier than 120 might be impossible with hot fusion. Detecting them will be equally hard: If the expected lifetimes drop too low, the atoms might not survive the 1-microsecond trip through the separator. They could decay midflight instead—ghost atoms that disappear without a trace.

Moving beyond 120, then, will probably require new approaches. "Multinucleon transfer reactions" would involve firing, say, uranium onto curium at relatively low speeds—another "soft touching". Their nuclei wouldn't fuse completely, but a chunk of one might break off and glom onto the other. Depending on the size of the chunk, scientists might even leap to much higher element numbers instead of inching along one atomic number at a time.

Such methods remain unproven, however. "Heavy-element scientists like to work one piece at a time," says Jacklyn Gates, leader of the heavy-element group at Lawrence Berkeley National Laboratory in California. And much beyond 120, she says, "We don't know enough to even know what to look for—what half-life to look for, what decay properties to look for."

Given those difficulties, some scientists propose ditching accelerators. In one approach, low-power nuclear blasts would induce fusion reactions in target atoms. That isn't as crazy as it sounds: Elements 99 and 100 were first identified in the fallout of atmospheric atomic bomb tests. Still, most scientists are skeptical of that approach given the obvious radiation hazards and the short lifetimes of superheavy atoms, which might expire before they could be sifted from the nuclear debris.

Other scientists suggest finding new elements the old-fashioned way: by hunting for them in nature. That was actually a popular pastime a few decades ago, as physicists scoured cosmic rays, meteorites, moon rocks, and even ancient shark teeth for superheavies. Nothing ever came of those projects. Nowadays, focus has shifted to supernova explosions and anomalous stars such as Przybylski's star, whose spectrum shows signs of einsteinium, which is otherwise never found in nature. Perhaps the star's hot, dense interior houses even heavier elements.

Still, there's no guarantee superheavy elements exist in nature. And the long dry spell—no new elements have been created since 2010—worries some researchers.

"If you look backwards over several decades," says Pekka Pyykkö, a theoretical chemist at the University of Helsinki, "people have made roughly one new element maybe every 3 years—until now." Today's barrenness could be the new normal.

2. Filling out the table

In recent decades, labs in several countries have extended the table with artificial elements. More might be created (gray) out to 172. Calculations by chemist Pekka Pyykkö of the University of Helsinki predict the table's periodicity will break down in certain places, leaving elements such as 139 and 140 out of numerical order.

Even if scientists can overcome the technical challenge of creating new elements, other questions remain: How many elements can exist, even hypothetically? How far could the periodic table go?

One prominent theory predicts an end at element 172. No one knows what will happen above that point, but for quantum mechanical reasons, an atom's nucleus might start to gobble up electrons and fuse them with protons, producing neutrons as a by-product. That process would continue until the proton count dropped back to 172, providing a hard cap on the atomic number. (And if that sounds weird, well, that's quantum mechanics.)

Other research suggests elements will run out long before 172. As nuclei get larger, the repulsive force between protons becomes overwhelming. By general consensus, a nucleus must survive for at least 10~14 seconds to count as a new element. Given how fragile elements in the 110s already are, heavier elements might struggle to hold on even that long. Some scientists predict that nuclei can overcome that problem by twisting into exotic shapes—hollow bubbles or even latticelike buckyballs. But other scientists doubt those shapes would be stable.

Which is a shame, because exciting things could happen in the 130s or 140s. In particular, the sine qua non of the periodic table—its periodicity—could break down completely.

In general, all elements within the same column of the table have similar chemical and physical properties. But that trend might not hold true forever. Scientists across the world have managed to probe the properties of single superheavy atoms by studying how they adhere to different materials. And the association between columns and chemical behavior already seems to be breaking down in the 110s.

Element 114, for instance, acts like a gas at room temperature, even though the element above it, lead, is about the most un-gas-like substance imaginable. Similarly, although element 118 falls into the noble gas column, theory predicts that it will readily attract electrons—something no other noble gas does. Those anomalies arise because of relativistic effects: The high, concentrated charge of a superheavy nucleus distorts the orbits of surrounding electrons, which affects how they behave and form bonds.

As Haba says, "The chemical properties of superheavy elements are very unique, and we cannot simply extrapolate." And although 114 and 118 seem to depart only modestly from expectations, even heavier elements could have wildly unexpected properties because relativistic effects will only grow larger as elements gain weight. So where should anomalous elements go? In the column where their atomic numbers say they should go or in a column with elements of similar properties?

The answer depends on whom you ask. For some scientists, the table is primarily about underlying atomic structure, not chemical behavior. Deviations are therefore not allowed. Other researchers are more pragmatic. "The periodic table is more useful for telling me what the chemistry of an element is, so I would argue for changing it around," Gates says.

Pyykkö has pushed the idea of anomalous elements to its extreme, calculating theoretical properties for all elements through 172 and arranging them into a futuristic table. The result is jarring: At one point, the sequence of atomic numbers jumps backward from 164 to 139 and 140 before skipping ahead to 169 (see table, left). The bizarro table now hangs on his office wall. "When I give talks," he says, "I usually joke that this periodic table should be enough for the rest of this century."

Beyond the divisions over the structure of the table, a deeper rift exists between people who think pursuing new elements is worthwhile and those who think it's a waste of time and resources. Gates voices her skepticism: "For elements 119 or 120, with our current technology, you're looking at years of beam time potentially for one atom—and what does that tell you?"

Still, she understands why some labs pursue new elements: "A new element is what makes people interested. And it does help you get funding. I just don't think it's science that's driving the experiments. It's politics." Indeed, RIKEN's 9-year pursuit of element 113 resulted in a nice budget boost. And because 113 was the first element created in Asia, the scientists became folk heroes in Japan. Someone even published a manga comic book about their work.

Dubna scientists argue their work is not mere trophy hunting. Karpov—who owns four sports jackets and wears a different Russian-themed element lapel pin on each (dubnium, flerovium, moscovium, and oganesson)—says making new elements can verify theoretical predictions about their half-lives and other properties.

He and his colleagues will also try, during some experimental runs, to add neutrons to existing superheavy elements and produce longer-lived versions of them. Nazarewicz, skeptical of making new elements, sees the value in that. "I would like us to get more stable," he says. Tinkering with existing elements might even allow scientists to reach the island of stability—a supposed region of longer-lived superheavy elements—and study those elements' properties. If nothing else, the technologies used to make new elements can help produce radioisotopes for medicine and test how well satellite components withstand bombardment by particles.

Ultimately, though, the search for new elements is its own reward—l'art pour l'art. "There's a majesty to increasing the number of protons," Karpov says. "It's natural to come to a limit" and try to push beyond. Plus, he says with a smile, his moscovium lapel pin gleaming, "sometimes it is good to say you did something first."

Unit 17　The Importance of Synthetic Chemistry in the Pharmaceutical Industry

Innovations in synthetic chemistry have enabled the discovery of many breakthrough therapies that have improved human health over the past century. In the face of increasing challenges in the pharmaceutical sector, continued innovation in chemistry is required to drive the discovery of the next wave of medicines. Novel synthetic methods not only unlock access to previously unattainable chemical matter, but also inspire new concepts as to how we design and build chemical matter. We identify some of the most important recent advances in synthetic chemistry as well as opportunities at the interface with partner disciplines that are poised to transform the practice of drug discovery and development.

Over the past century, innovations in synthetic chemistry have greatly enabled the discovery and development of important life-changing medicines, improving the health of patients worldwide. In recent years, many pharmaceutical companies have chosen to reduce their R&D investment in chemistry, viewing synthetic chemistry more as a mature technology and less as a driver of innovation in drug discovery. Contrary to this opinion, we believe that excellence and innovation in synthetic chemistry continues to be critical to success in all phases of drug discovery and development. Moreover, recent developments in new synthetic methods, biocatalysis, chemoinformatics, and reaction miniaturization have the power to accelerate the pace and improve the quality of products in pharmaceutical research. The application of new synthetic methods is rapidly expanding the realm of accessible chemical matter for modulating a broader array of biological targets, and there is a growing recognition that innovations in synthetic chemistry are changing the practice of drug discovery. Here, we identify some of the most enabling recent advances in synthetic chemistry as well as opportunities that we believe are poised to transform the practice of drug discovery and development in the coming years.

The pharmaceutical sector is currently facing multiple challenges: an increasing focus on complex diseases with unknown causal biology, a rapidly changing and highly competitive land-scape, and substantial pricing pressures from patients and payers. In this challenging environment, drug discovery scientists must select biological targets of relevance to human disease and safe and effective therapeutic molecules that appropriately modulate those targets. The current toolbox of synthetic methods and com-mon chemical starting materials provides access to chemical space that can be efficiently explored and mined to identify a suitable ligand and subsequently pursue studies of that preliminary lead compound toward its potential development as a successful drug. Brown and Boström have noted that a historical overreliance on just a few robust synthetic transformations (amide bond formation, sp^2-sp^2 C-C cross-coupling, and SNAr reactions) has biased the output of many drug discovery efforts, leading to narrow sampling of chemical space. In other cases, the lack of any reasonable method of synthesis has, at mini-mum, hampered thorough evaluation of chemical space or, at worst, prevented it completely.

Conversely, the discovery of breakthrough synthetic methods can truly transform the process of drug discovery. Innovation in synthetic chemistry provides opportunity to gain more rapid access to biologically active, complex molecular structures in a cost-effective manner that can change the practice of medicine. An outstanding example of the transformative power of synthetic chemistry in drug discovery is the application of carbenoid N-H insertion chemistry to the syn thesis of β-lactam antibiotics. In the 1950s, the synthesis of antibiotics such as penicillin represented a formidable challenge to medicinal chemists, and broad exploration of structure-activity relationships (SAR) within this class of compounds was hindered by a lack of good methods of synthesis for these chemically sensitive structures. Indeed, the first chemical synthesis of penicillin took nearly a decade of dedicated effort to achieve despite an intensive effort across multiple laboratories. This lack of synthetic accessibility prevented thorough evaluation of structurally related antibiotics that might have a broader spectrum of activity and an improved resistance profile. The application of intramolecular N-H carbenoid insertion chemistry (Fig. 17-1) to these structures provided a disruptive solution to the preparation of these fused β-lactams. This synthetic method was applied to the preparation of numerous natural and synthetic anti-infectives, including thienamycin, which subsequently led to the discovery and industrial manufacture of the antibiotic imipenem. In this example, synthesis enabled design, opening access to previously unattainable molecules of high therapeutic value.

The development of targeted medicines for the treatment of chronic hepatitis C infection, a global health challenge, illustrates another key advance that innovative synthetic chemistry has contributed to drug discovery in recent years. The design and synthesis of hepatitis C virus (HCV) NS3/4a protease inhibitors represents a formidable challenge for medicinal chemists because the active site of this protease has a shallow, open binding site, and the enzyme possesses both genotypic and mutational diversity. Early studies of peptide-based inhibitors and subsequent molecular modeling suggested that construction of large, macrocyclic enzyme inhibitors could provide favorable ligand-protein binding and potent inhibition of this essential viral pro-tease. The relatively flat and featureless protein surface requires a large ligand to gain sufficient binding affinity, while constrained macrocyclic ligands minimize the entropic cost of inhibitor binding. The application of ring-closing metathesis chemistry has been trans-formative in the synthesis of many HCV NS3/4a protease inhibitors of varying ring sizes and complexity, including six approved drugs: simeprevir, paritaprevir, vaniprevir, grazoprevir, voxilaprevir, and glecaprevir. Ring-closing metathesis chemistry enabled the discovery of these and related macrocycles, allowing rapid assembly of complex bioactive molecules and broad exploration of SAR to address a range of properties.

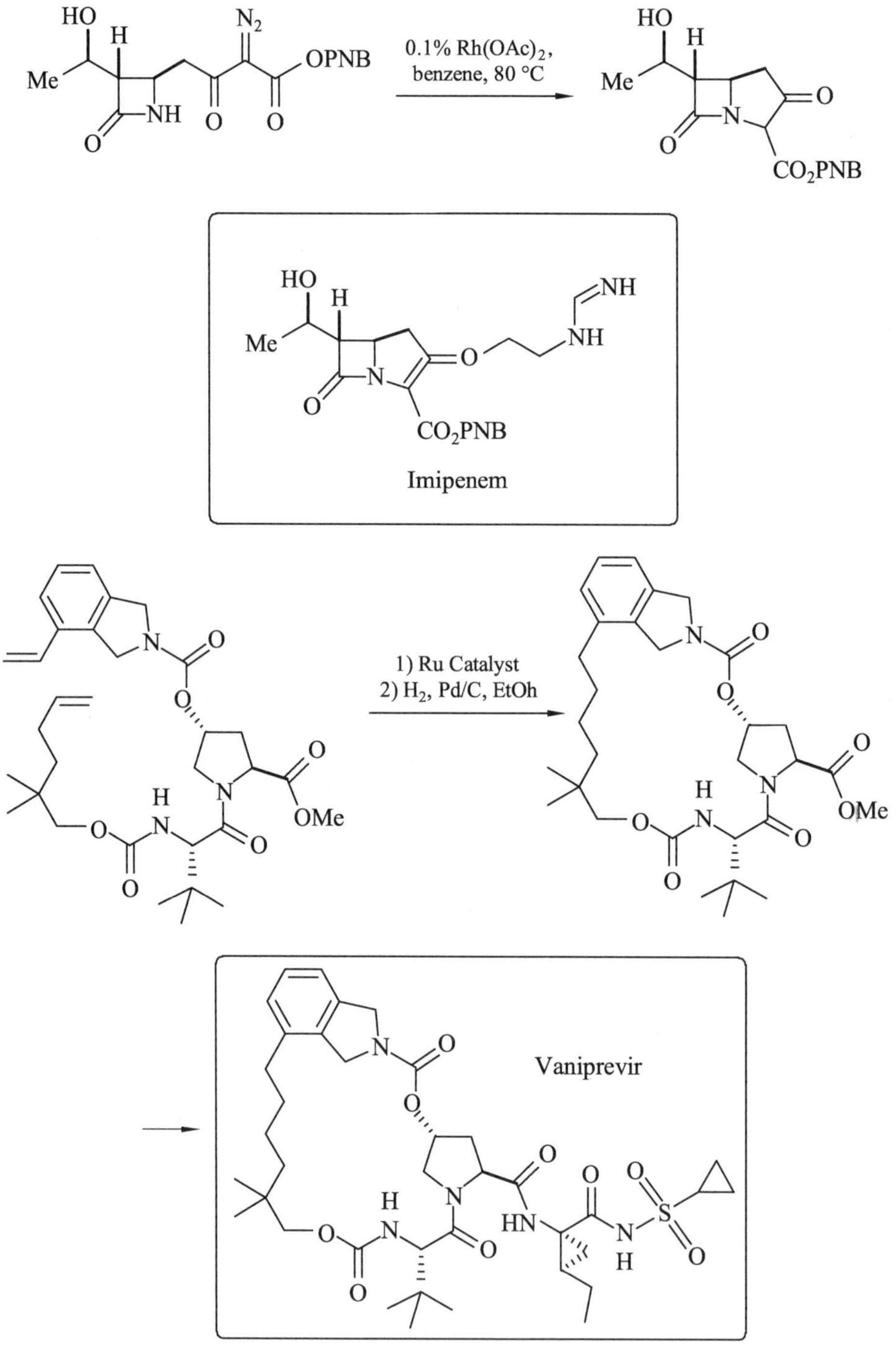

Fig. 17-1 Synthetic method innovations enable discovery of important anti-infectives, imipenem and vaniprevir.

In the two examples described above, the discovery of new synthetic pathways changed the way scientists thought about designing and building molecules, which broadened the accessible chemical space and thereby furnished molecules possessing the biological activity required in future drug candidates. The ability of the pharmaceutical industry to discover molecules to treat unmet medical needs and deliver them to patients efficiently in the face of an increasingly challenging regulatory landscape is dependent on continued invention of transformative, synthetic methodologies. Toward this end, investment in research directed toward synthetic methods innovation, furthering the nexus of synthetic chemistry and biomolecules, and developing new technologies to accelerate methods discovery is absolutely essential. Pertinent examples in these three areas are reviewed below.

1. Synthetic methods innovation

Over the past 20 years, several scientists have been recognized with the Nobel Prize for the invention of synthetic methodologies that have changed the way chemists design and build molecules. Each of these privileged methods—asymmetric hydrogenation, asymmetric epoxidation, olefin metathesis, and Pd-catalyzed cross-couplings—have broadly influenced the entire field of synthetic chemistry, but they have also enabled new directions in medicinal chemistry research. Of particular interest are new synthetic methods that enable medicinal chemists to control reactivity in complex, drug-like molecules, access non-obvious vectors for SAR development, and rapidly access new chemical space or unique bond formations. Recently, there have been several reported methods in these categories that have been rapidly adopted by medicinal chemists as a result of their practicality and broad utility.

Owing to the diverse biological activity of nitrogen-containing compounds, the discovery of Pd-catalyzed and Cu-catalyzed cross-coupling reactions of amines and aryl halides to form C-N bonds resulted in the rapid implementation of these synthetic methods in the pharmaceutical industry. The methodology addressed an unsolved problem to quickly and predictably access aromatic and heteroaromatic amines from simple precursors, and as a result it was rapidly adopted by medicinal chemists. Further development of these methodologies by process chemistry groups for scale-up has resulted in optimized ligands and precatalysts, as well as generally reliable protocols that have further advanced the application of this methodology in discovery programs. Consequently, aromatic C—N bonds are common features in pharmaceutical compounds, highlighting the tremendous impact that controlled construction of C—N bonds in aromatic compounds has had on medicinal chemistry programs. The next frontier is the development of reliable methods to accomplish C sp^3-N couplings.

As the development of transition metal-catalyzed processes has advanced, application of cutting-edge methods to the predictable activation of C—H bonds for functionalization of complex lead structures can enable novel vector elaborations, changing the way analogs are prepared. In particular, late-stage selective fluorination and trifluoromethylation of C—H bonds in an efficient, high-yielding, and predictable fashion permits the modification of lead compounds to give analogs that potentially possess greater target affinity and metabolic stability without resorting to de novo synthesis. Methodological advances have enabled preparation of fluorinated analogs of lead structures under either nucleophilic or electrophilic conditions. One promising recent example shows that electrophilic aromatic fluorination can occur under mild conditions with a palladium catalyst and an electrophilic fluorine source such as N-fluorobenzenesulfonimide (NFSI). In addition, trifluoromethylation of a structurally diverse array of drug discovery candidates using zinc sulfinates, in the presence of iron(III) acetylacetonate, generated analogs with improved metabolic properties. Visible-light photoredox catalysis has been also been applied to the practical, direct trifluoromethylation of het-eroarenes.

Adoption of photoredox catalysis in the pharmaceutical industry has been rapid, owing to the practicality of the process, the tolerance to functional groups in drug-like candidates, and the activation of nonconventional bonds in drug-like molecules. Application of photoredox catalysis to the Minisci reaction was reported, enabling the facile and selective introduction of small alkyl groups into a variety of biologically active heterocycles such as camptothecin. Photoredox catalysis has also been used for the direct and selective fluorination of leucine methyl ester to afford g-fluoroleucine methyl ester with a decatungstate photocatalyst and NFSI (Fig. 17-2). Numerous processes have been reported to access g-fluoroleucine methyl ester, a critical fragment of the late-stage drug candidate odanacatib; however, this method enables the most direct and efficient method to access this key building block in the fewest operations from a commodity feed-stock. More recently, photoredox catalysis was used to generate diazomethyl radicals, equivalents of carbyne species, which induced site-selective aromatic functionalization in a diverse array of drug-like molecules. This represents the latest of a series of very diverse, practical, and potentially impactful uses of photoredox tech-niques to assemble libraries of drug-like scaffolds for screening.

Novel vector elaboration: C-H activation/functionalization

$Na_4W_{10}O_{32}$(cat.) / NFSI, CH_3CN-H_2O / 365 nm $h\nu$

Although the preceding examples highlight the power of photoredox catalysis to accomplish previously unimaginable reactivity under very mild conditions, even more remarkable transformations are being reported via synergistic catalysis, where both the photocatalyst and a cocatalyst are responsible for distinct steps in a mechanistic pathway that is only accessible with both catalysts present. For example, the combination of single-electron transfer-based decarboxylation with nickel-activated electro-philes has provided a general method for the cross-coupling of sp^2-sp^3 and sp^3—sp^3 bonds. This method establishes a new way of thinking about the carboxylic acid functional group as a masked cross-coupling precursor, expanding the synthetic opportunities for a functional group that is ubiquitous in chemical feedstocks. Furthermore, leveraging synergistic catalysis with photoredox has resulted in the discovery of milder conditions for C-O and C-N cross-couplings, allowing application of these methods to more pharmaceutically relevant substrates. The concise synthesis of the antiplatelet drug tirofiban is an excellent example of how the pharmaceutical industry can readily use this methodology to facilitate drug discovery and development. As research continues to surge in this field, additional breakthroughs are anticipated, and these will likely change how molecules are designed and built.

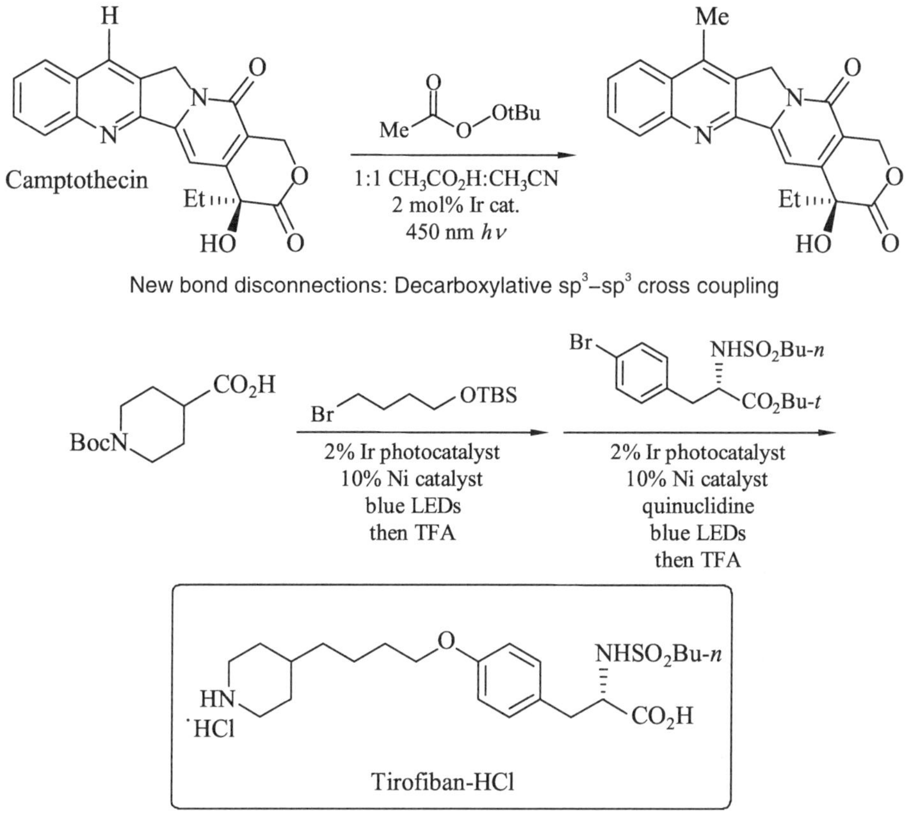

Fig. 17-2 Synthetic methods with potential to enable drug discovery.

2. Intersection of synthetic chemistry with biomolecules

Biopolymers including proteins, nucleic acids, and glycans have evolved to achieve exquisite selectivity and function in a highly complex environment. These properties are of great interest to the pharmaceutical industry not only from a target perspective, but also from a therapeutic perspective. The success of monoclonal antibodies, peptides, and RNA-based therapies attests to the power that nature's platforms offer to our industry and patients. Recent advances in merging the fields of synthetic and biosynthetic chemistry have sought to harness these molecules and to expand useful manipulation of biomolecules in three distinct ways: as catalysts for novel and selective transformations, as conjugates through innovative bio-orthogonal chemistry, and in the development of novel and improved therapeutic modalities.

3. Biocatalysis

Historically, the broad adoption of biocatalysis was held back by a limited availability of robust enzymes, a relatively small scope of reactions, and the long lead time required to optimize a biocatalyst through protein engineering. The invention of a recombinant engineered Merck/Codexis transaminase biocatalyst for the commercial manufacture of sitagliptin (Januvia) has inspired the broader application of bio-transformations in the pharmaceutical industry. Tremendous advances have been made in molecular biology, bioinformatics, and protein engineering, enabling the development of biocatalysts with desired stability, activity, and exquisite selectivity. The impact of this area of research is exemplified by the 2018 Nobel Prize in Chemistry, recognizing Frances Arnold "for the directed evolution of enzymes". As a result, biocatalysis has become more prevalent as a tool in drug discovery, as a valuable method for drug metabolite synthesis, and as a tool to enable rapid analog synthesis for SAR. For example, in 2013, the important discovery that cyclic guanosine monophosphate-adenosine monophosphate (2′,3′-cGAMP) is the endogenous agonist of STING, a protein involved in the activation of innate immune cells, triggered an intense interest in the synthesis of cyclic dinucleotide (CDN) analogs. Typically, the total synthesis of CDNs by purely chemical transformations requires long linear sequences and results in a time-consuming and low-yielding process. The optimization of STING agonists was greatly facilitated by the realization that the endogenous enzyme cGAS, responsible for the in vivo production of 2′,3′-cGAMP, could be engineered and harnessed for the biocatalytic production of non-natural CDNs (Fig. 17-3). The cyclization of various nucleotide triphosphate derivatives in a single biosynthetic step considerably reduced the cycle time and increased the yield of CDN synthesis, inspiring the design of novel agonists and the generation of SAR in this class. The continued investment in biocatalysis will lead to innovative solutions for unsolved problems in synthetic chemistry in both the discovery and development arenas. This will be driven by increased speed of protein engineering, access to enzymes with a variety of natural and even unnatural catalytic activities, and the implementation of biocatalytic cascade catalysis to efficiently build complex chemical matter from simple starting materials.

2′, 2′-cGAMP

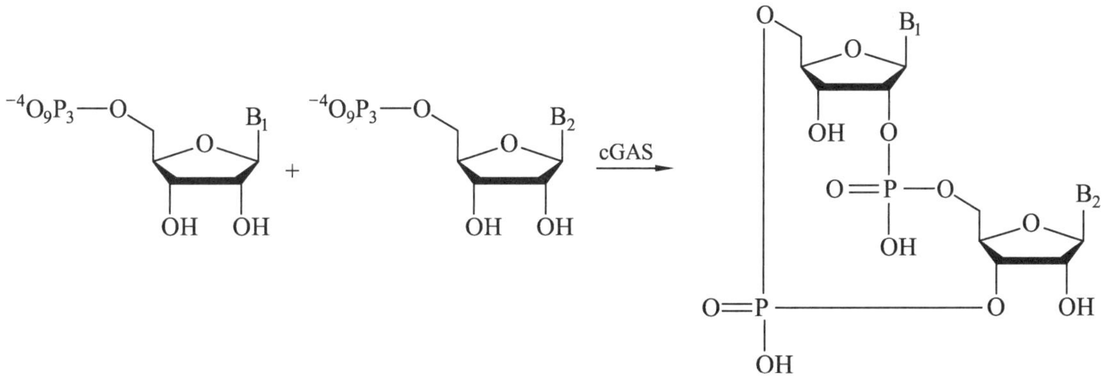

Fig. 17-3　Biocatalytic synthesis of novel cyclic dinucleotides.

4. Bio-orthogonal chemistry

Achieving selective reactions with biopolymers such as proteins presents a host of unique challenges to the synthetic chemistry community; proteins have multiple reactive centers, charged residues, higher-order structure, and are usually handled in an aqueous environment. Nonetheless, the opportunity to create improved conjugates as therapies and imaging agents, or to induce covalent interactions to identify protein targets, represents important value to therapeutic drug discovery.

Methods for selective conjugation to biomolecules have undergone major synthetic evolution over the past 20 years. The discovery and development of a suite of click reactions has served as a powerful and broadly applied tool in protein bioconjugation. This highly bio-orthogonal and biocompatible reaction offers a powerful alternative to heterogeneous conjugation to surface lysines or engineered cysteines, and spurred the development of complementary expression technologies that could incorporate unnatural elements or recognition tags into biopolymers. This evolution in conjugation chemistry is best evidenced in the field of antibody-drug conjugates (ADCs): The first generation of ADCs were heterogeneous conjugates, whereas those of the second generation are now almost entirely homogeneous, with growing evidence that the site of conjugation is an important determinant of overall ADC performance.

The development of additional bio-orthogonal chemistries that can lead to selective reaction with biomolecules, particularly without the requirement for engineering a recognition element into the biomolecule, is an important new frontier for synthetic impact. Two recent examples of synthetic innovation suggest this toolset is expanding for proteins. In many cases, having the ability to conjugate at either the N or C terminus of a wild-type protein should avoid un-intended disruption of its function or secondary structure. The development of selective N-terminal conjugation chemistry and complementary application of decarboxylative alkylation chemistry to the C terminus of a protein substrate offer new insights into achieving bio-orthogonal and highly site-selective conjugation with complex biomolecules (Fig.17-4). These reactions take advantage of local differences in basicity and ionization potential respectively and, in doing so, leverage the complexity that biopolymers offer.

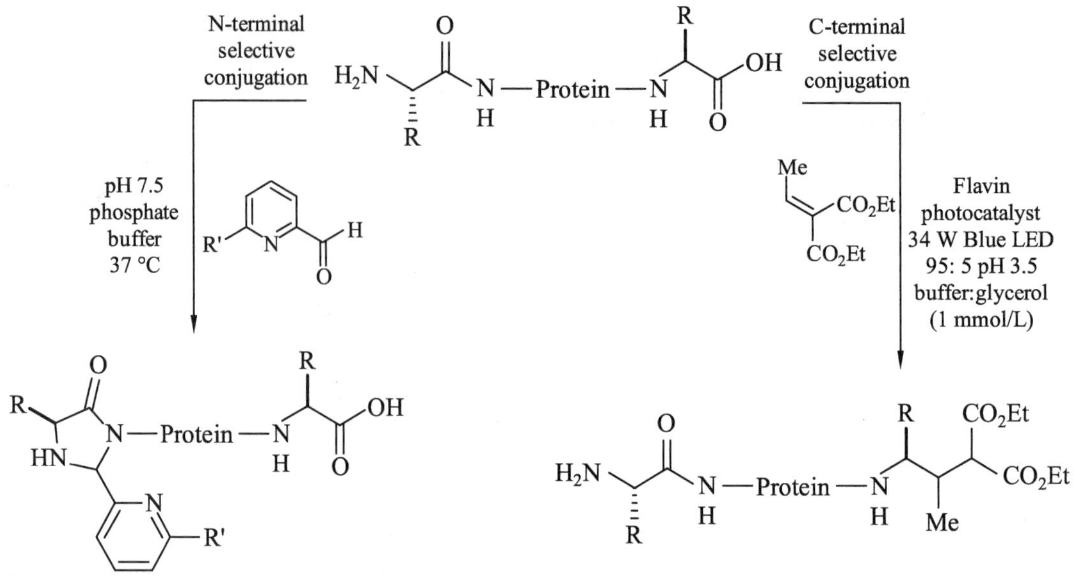

Fig. 17-4 Bio-orthogonal reactivity with proteins at N and C termini.

5. Synthetic innovation and therapeutic modalities

As these advances in synthetic, biorthogonal, and biosynthetic chemistry merge, so too do our capabilities to improve therapeutic modalities in the space between synthetic small molecules and expressed large monoclonal antibodies. Peptides, oligonucleotides, and bioconjugates have been advanced particularly for biological targets deemed "undruggable" by small-molecule and antibody platforms. Advances in these chemistries inspire new platforms and improve the breadth of biological targets that we can address. Two examples of innovation in therapeutic modalities through synthetic and biosynthetic chemistry are described below, although many others are being invented in academic and industrial settings.

In the first case, it has long been appreciated that a critical element of the success of oligonucleotide-based therapies was the introduction of phosphorothioates into the oligo backbone, which afforded improved stability to biological matrices as well as improved membrane permeability to aid with cytosolic delivery. Although these and other improvements in stability and delivery have advanced the field and enabled novel therapeutics to enter the clinic, many oligo-based therapies require high doses to overcome barriers to delivery, and their use is limited by their toxicity. Further improvements in stability and potency of the oligonucleotide should contribute to a widening of the therapeutic index and dose lowering. Interestingly, the chemistry used to introduce stabilizing phosphorothioates leaves each center as a mixture of two P-stereoisomers. Therefore, most clinical phosphorothioate-containing oligos that have 20 base pairs are, in reality, a large mixture of stereoisomers (219), each with different potency and stability characteristics. The ability to control phosphorothioate chemistry through an oxazaphospholidine approach by Wada and colleagues led to a practical and scalable platform for stereopure antisense oligonucleotides that demonstrate preclinical superiority to the corresponding stereomixtures.

Within the peptide arena, there has been a growing recognition that cyclic peptides offer improved starting points for drug discovery programs relative to their linear counterparts, largely due to improvements in entropic cost for binding and proteolytic stability. Early display platforms developed to discover cyclic peptides relied on disulfide formation, and more recently on post-translational introduction of bis-electrophiles that can cyclize peptides with two cysteine residues. Through combined application of a ribozyme biocatalyst to enable unnatural amino acid incorporation into peptides, and then bio-orthogonal chemistry for cysteine cyclization through that unnatural amino acid, the Suga lab has developed an improved mRNA display platform that has demonstrated tremendous potential to identify peptide ligands for challenging targets. The merging of chemical synthesis and biosynthesis within a common platform inspires further exploration of cyclic peptide modality; the introduction of selection pressures and forced evolution into this platform begins to resemble aspects of natural product generation that has historically inspired both organic synthesis and drug discovery.

6. Technologies to accelerate innovation high-throughput experimentation

Given the need to invent and rapidly deliver medicines to patients, the pharmaceutical industry must invest in capabilities with the potential to radically accelerate the discovery and industrialization of transformative synthetic methodologies. High-throughput screening in biology has been the foundation of hit discovery for decades, and in recent years, the pharmaceutical industry has strategically invested in the creation of high-throughput experimentation (HTE) tools for chemistry that enable scientists to test experimental hypotheses with hundreds of arrayed experiments. In the same time frame required for traditional single-reaction evaluation, the different parameters that determine reaction outcome, discrete variables (catalysts, reagents, solvents, additives), and continuous variables (temperatures, concentrations, stoichiometries) can be holistically explored in parallel. As a result, the synthetic chemist now has access to exponentially larger amounts of experimental data than ever before. One recent example of the use of end-to-end HTE in process development was the discovery and development of an organo-catalyzed, enantioselective, aza-Michael reaction for the commercial manufacture of the antiviral letermovir (Fig.17-5). In this work, a series of efficient synthetic pathways were envisioned by chemists and key transformations were evaluated inparallel using HTE. The emergence of an H-bonding catalysis mechanism was initially discovered with moderate enantioselectivity and low conversion using chiral phosphoric acids. Rapid evaluation of a large number of diverse scaffolds with H-bonding capability in this transformation resulted in the discovery of an efficient and highly selective bis-sulfonamide catalyst. Further HTE work enabled the mechanistic understanding of the transformation, leading to optimization of both the catalyst structure and definition of optimal processing conditions. In this study and in many others, novel bond-forming reactions were conceived by scientists, discovered through HTE, and then rapidly industrialized for the commercial manufacture of late-stage drug candidates.

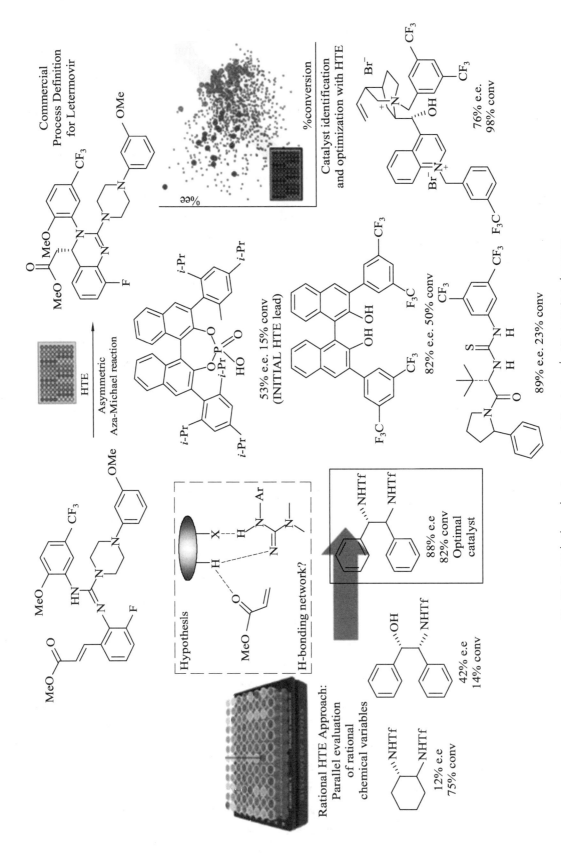

Fig. 17-5　High-throughput experimentation to accelerating reaction discovery.

HTE tools have also begun to have an impact in drug discovery. As new catalytic methods emerge that redefine which bonds can be forged, the breadth of the resulting substrate scope is poorly understood, as most test substrates commonly demonstrated in the literature are simple and not representative of the complex functionality common in drug candidates. Pre-dosed, reaction-specific HTE screening kits, containing a lab's most successful and general catalyst systems, are used in discovery chemistry labs to enable the rapid identification of reaction conditions that work for these complex substrates. Additionally, HTE has recently been leveraged to benchmark emerging methods against different catalytic procedures through the creation of arrays of complex, drug-like substrates known as informer libraries or through addition of diverse molecular fragments that can disrupt catalysis. The use of these diagnostic methods allows exploration of the relationship between reaction types and diverse complex substrate structures, thus enabling synthetic practitioners to make better decisions about which synthetic methods to prioritize in their problem-solving. Additionally, miniaturization of HTE to nanomole scale—for example, by automated nanomole-scale batch and flow approaches—now enables the execution of more than 1,500 simultaneous experiments at microgram scale in 1 day for rapid identification of suitable reaction conditions to explore chemical space and accelerate drug discovery. This capability is augmented by advances in rapid high-throughput analytics, such as MISER (multiple injections in a single experimental run) and MALDI (matrix-assisted laser desorption/ionization) mass spectrometry techniques, which have enabled the analysis of as many as 1,536 reactions in very short time frames. Finally, nanomole HTE can also expedite the preparation of diverse, complex arrays of molecules and, when coupled directly with biological testing, can radically alter how drug discovery is performed.

7. Computational methods

The use of computer-assisted methods to guide synthetic chemistry is emerging as an important component in the practice of drug discovery. Advances in computational chemistry and machine learning in the past decade are delivering real impact in areas such as new catalyst design or showing considerable promise in others such as reaction prediction. The application of deep learning methods has the potential to uncover new chemical reactions, expanding the access to new pharmaceutical chemical matter. Granda etal. have reported promising results toward this end. By combining automated synthesis with machine learning, they reported the discovery of four chemical transformations with differentiated novelty.

Recently, computer-guided design has been successfully applied to the preparation of catalysts that provide asymmetric control of a cy-cloisomerization reaction. Computational methods were used to evaluate the catalytic pathway of a previously unknown reaction, leading to the hypothesis that the electronics of the catalyst ligand influence both the rate and stereo-selectivity of the transformation. Application of quantum methods such as density functional theory (DFT) provided optimal ligand designs with markedly enhanced rate and selectivity over the original ligand. A second example where the use of computational methods aided in the design of a superior catalyst is reported in the synthesis of a pronucleotide (ProTide, Fig.17-6). Achieving selective

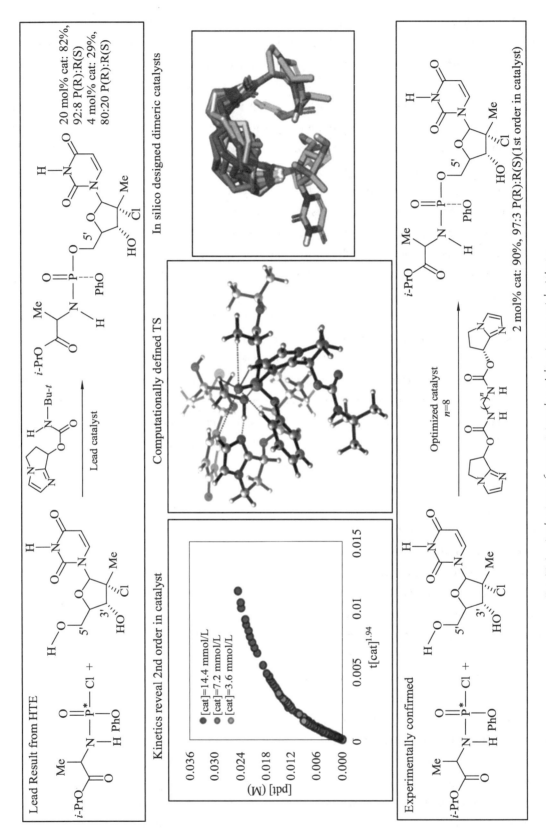

Fig. 17-6 Application of computational modeling to new catalyst design.

phosphoramidation of a nucleoside at the 5' hydroxyl lover the 3' hydroxyl with stereocontrol at the phosphorus center is highly challenging. A combination of mechanistic studies using a variety of chiral catalysts and DFT calculations of a proposed transition state further informed by experimental observations led to the rational design of a dimeric phosphoramidation catalyst with an improved rate and excellent stereoselectivity.

Despite these successes, the process for rational computational design of acatalyst is arduous, requiring the modeling of multiple mechanistic pathways and refinement of numerous molecules and transition states. A program for automating much of this process has been reported, and the advancement of such methods as well as the continual increase in processing power will drive further use of these tools in the future.

The application of machine learning to synthetic problems has also generated considerable interest and excitement. One area of active research is the use of algorithms for synthetic route planning to a target molecule. Segler et al. combined Monte Carlo tree search and three neural networks to identify potential synthetic routes. The success of the approach was qualitatively evaluated through a double-blind A/B test, where 45 chemistry students showed no preference between machine-suggested synthetic routes versus literature routes for representative target molecules. Machine learning has additionally been applied to forward reaction prediction. Neural networks were used to predict the major product of a reaction using an algorithm that assigns a probability and rank to potential products. Additionally, machine learning was used to successfully predict the performance of a single reaction, a Buchwald-Hartwig amination, against multiple variables: reactants, catalysts, bases, and additives. Application of machine learning holds considerable promise for synthetic optimization of targets far exceeding those described herein, toward predicting routes, main products, side products, and optimal conditions, among others. The continued advancement of these methods leverages the wealth of public information in the scientific and patent literature as well as within pharmaceutical institutions. The quality, breadth, depth, and density of the data within the domain of the predictions is critical for driving toward high-accuracy models. Inclusion of examples of both successful and unsuccessful transformations is also highly important. HTE is a highly attractive, complementary technology for augmenting existing datasets by generating model-suitable data, maximizing information content through careful design of experiments and capacity to deliver large volumes of data in a rapid and cost-effective manner.

8. Future directions

As we have discussed, breakthroughs in synthetic chemistry have proven to be the inspiration for the discovery and development of new medicines of important therapeutic value. Despite the many advances described above, the pace and breadth of molecule design is still constrained because of unsolved problems in synthetic chemistry. Many opportunities still remain to advance the field, such that synthetic chemistry will never constrain compound design or program pace, and should actually inspire access to uncharted chemical space in the pharmaceutical industry.

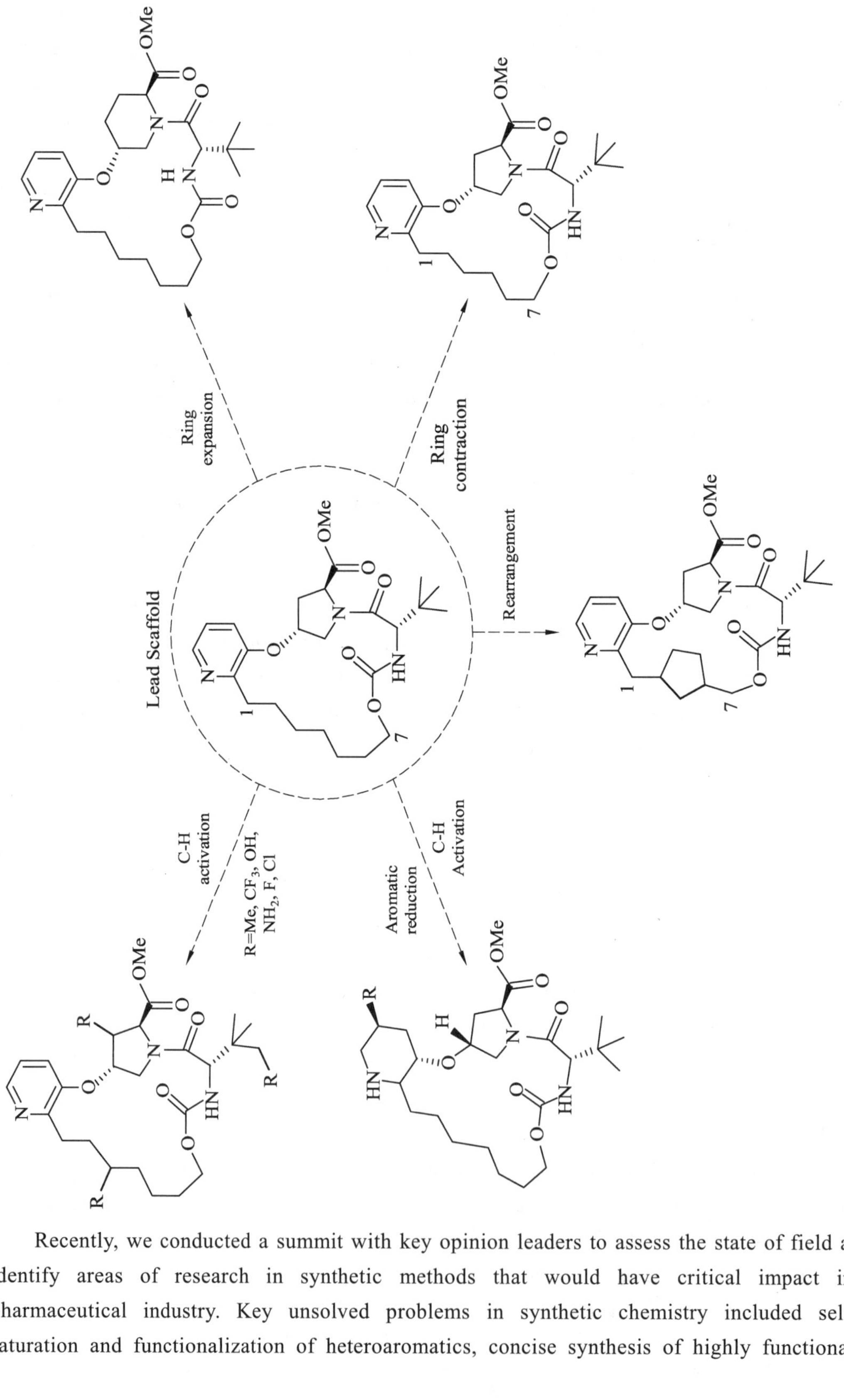

Fig. 17-7 Molecular editing to enable drug discovery.

Recently, we conducted a summit with key opinion leaders to assess the state of field and to identify areas of research in synthetic methods that would have critical impact in the pharmaceutical industry. Key unsolved problems in synthetic chemistry included selective saturation and functionalization of heteroaromatics, concise synthesis of highly functionalized, constrained bicyclic amines, and C-H functionalization for the synthesis of α,α,α-trisubstituted amines. Other areas, such as selective functionalization of bio-molecules and synthesis of noncanonical nucleo-sides, were identified as emerging areas of high potential impact. We envision that partnerships between the pharmaceutical industry and leading academic groups in the field hold great promise to spur the invention of disruptive synthetic chemistry to address these areas.

The most intriguing idea to emerge from the discussion was the concept of molecular editing, which would entail insertion, deletion, or exchange of atoms in highly functionalized compounds at will and in a highly specific fashion. Many innovations discussed above possess elements of this aspirational goal; however, a truly general method of this type would substantially change the pace of drug discovery and reduce constraints on compound design. Figure 17-7 prospectively illustrates how analogs of a complex lead scaffold might be accessed via site selective C-H functionalization, heteroaromatic reduction, ring expansion, or ring contraction. The power to modify this scaffold directly and specifically not only avoids a potentially lengthy synthesis of analogs, but also removes any limitation of molecular design imposed by synthetic hurdles. We anticipate that breakthroughs in the area of molecular editing will improve the pace and quality of molecule invention, enabling the introduction of new and important medicines at a faster rate.

9. Outlook

Synthetic chemistry has historically been a powerful force in the discovery of new medicines and is now poised to have an even greater impact to accelerate the pace of drug discovery and expand the reach of synthetic chemistry beyond the traditional boundaries of small-molecule synthesis. New methods of synthesis can greatly expand the rate of molecule generation while also providing opportunities to routinely synthesize complex molecules in the course of drug discovery. Manipulation of biomolecules either as catalytic reagents (i.e., engineered enzymes) or as substrates for site-specific modulation is becoming more accessible and creating new opportunities for producing novel therapeutic entities. Academic research continues to be an important venue for producing novel reactivity, and rapid application of new methods has the potential to further drive molecule invention in drug discovery. New technologies such as HTE, automation, and new analytical methods are accelerating the discovery of new reaction methods. Further, integration of computational reaction modeling with the vast quantities of experimental data generated by nanoscale HTE has the potential to build more informative models that can predict successful reaction conditions or even discover new reactions. The field of predictive chemical synthesis remains nascent, but opportunities to build prognostic algorithms via machine-learning processes are likely to expand in the coming years. Continued investment in synthetic chemistry and chemical technologies has the promise to advance the field closer to a state where exploration of chemical space is unconstrained by synthetic complexity and is only limited by the imagination of the chemist. Advancements in synthetic chemistry are certain to remain highly relevant to the mission of inventing new medicines to improve the lives of patients worldwide.

Unit 18   Extending the Application of Biocatalysis to Meet the Challenges of Drug Development

The pharmaceutical industry, driven by an increasing need to deliver new and more effective medicines to patients, is increasingly turning to the use of engineered biocatalysts for both lead generation of active compounds and the sustainable manufacture of active pharmaceutical ingredients. Advances in enzyme discovery, high-throughput screening and protein engineering have substantially expanded the available biocatalysts, and consequently, many more synthetic transformations are now possible. Enzymes can be fine-tuned for practical applications with greater speed and likelihood of success than before, thereby leading to greater predictability and confidence when scaling up these processes. Coupled with a greater awareness of which reactions are suitable for biocatalysis (for example, biocatalytic retrosynthesis), new chemoenzymatic and multi-enzyme processes have been designed and applied to the synthesis of a range of important pharmaceutical target molecules. Increasingly, researchers are exploring opportunities for using immobilized biocatalysts in flow conditions. In this Review, we discuss some of the key drivers and scientific developments that are expanding the application of biocatalysis in the pharmaceutical industry and highlight potential future developments that likely will continue to increase the impact of biocatalysis in drug development.

Our ability to discover and engineer new enzymes with expanded substrate scope and new reaction chemistries is now proceeding faster than ever before. Advances in enzyme discovery by, for example, bioinformatics, metagenomics, enzyme promiscuity and de novo design coupled with enzyme engineering (by directed evolution and high-throughput screening) have led to a rapid increase in the number of biocatalytic tools that are available for synthetic chemistry (Fig.18-1). When combined with a much greater awareness of the steps in target molecule synthesis for which biocatalysts can be used (biocatalytic retrosynthesis) (Fig.18-1), we are now clearly entering a golden age that offers unprecedented opportunities for the application of biocatalysis across a broad range of disciplines.

Biocatalysis has hitherto largely been viewed as a process technology (Fig.18-1). Process chemists increasingly consider the use of biocatalysis for developing scalable routes to target molecules, including second-generation manufacturing processes. Biocatalysis offers substantial benefits, including reduced cost of goods, number of synthetic steps and environmental impact, as well as improved safety and selectivity. A remarkable and attractive aspect of biocatalysis is the ability to rapidly generate novel enzymes that can then be optimized for a given substrate and reaction environment, enabling the engineering of biocatalysts to fit the process instead of vice versa. Consequently, a biocatalyst that is optimized for the synthesis of one pharmaceutical inter-mediate may never be used in the synthesis of another, and instead, a different enzyme will be evolved to more perfectly fit the requirements of the alternative synthetic scheme. This paradigm exemplifies the true power of biocatalysis and one that, in our experience, cannot yet be achieved using chemocatalysis. Furthermore, these engineered biocatalysts can now be easily arrayed in 96-well format, enabling rapid screening against substrates of interest to identify lead hits for further development and optimization. Examples of recently developed biocatalytic enzyme families with broad substrate scope include ketoreductases (KREDs) and transaminases.

The availability of a broader range of biocatalysts with an expanded substrate scope poses the question of whether an increased use of biocatalysts in discovery chemistry (Fig.18-1) to generate novel lead compounds for further optimization and to identify new ways of making target molecules is feasible. The use of immobilized biocatalysts in flow format could enable implementation of this approach (Fig.18-1). To effectively regulate biological activities and eliminate disease, we must be able to manipulate all forms of chemical matter (both large molecules and small molecules) that are involved in biological processes. As biocatalysis is the predominant way that signalling molecules are synthesized and biological processes are regulated in living organisms, we can use the same approach to more rapidly access the desired complex chemical space.

This shift in thinking opens up new opportunities for applying biocatalysis in areas in which existing chemistry is underdeveloped or inefficient, such as late-stage functionalization, C–H activation, reductive amination, halogenation and oligonucleotide synthesis. The use of biocatalysts also opens up new chemical space that is not accessible using alternative chemical reagents or catalysts. Biocatalysts are increasingly being used to create libraries of compounds for screening by parallel medicinal chemistry (PMC). In addition, multi-enzyme cascade processes enable the rapid introduction of chemical complexity into target molecules. Cascade reactions are increasingly being used in the pharmaceutical industry and account for an increasing number of enzyme-catalysed processes in development. Biocatalysis can make the use of protecting groups redundant, thereby removing additional wasteful steps from chemical synthesis. Indeed, an enzyme can be considered a combination of a catalyst and a directing or protecting group, as enzymes usually catalyse a transformation with exquisite regioselectivity.

Key issues in drug development include accelerating the process of screening biocatalysts and identifying potential hits for rapid scale-up and proof-of-concept studies. A number of developments are markedly improving the speed of biocatalyst hit identification, including those in modelling and bioinformatics, as well as the availability of a much wider range of new and evolved enzymes. High-throughput assays and data analysis are key factors here; for example, multiple injections in a single experimental run (MISER) chromatography can be used to assay thousands of reactions at a rate of one sample every few seconds. Moving from relatively simple statistical analyses of results to the use of machine learning algorithms requires the generation of very large, well-controlled data sets.

In this review, we highlight the increasing number of opportunities for applying biocatalysis in the pharmaceutical industry and discuss the challenges from both an academic and an industrial viewpoint. We provide an assessment of current trends and developments in the use of biocatalysis and also try to anticipate areas in which biocatalysis could have the most impact in the coming years.

Enzyme engineering
- Directed evolution
- Enzyme promiscuity

Enzyme discovery
- Computational design
- Bioinformatics (sequence-based discovery)
- Ancestral sequence reconstruction

Enzyme design features
- Catalytic activity
- Selectivity (regioselectivity, stereoselectivity, chemoselectivity)
- Specificity
- Stability

↓

Biocatalysts
- New classes of enzymes
- New reaction chemistries

↓

Medicinal chemistry and process chemistry
- Enzyme immobilization
- Flow chemistry

Organic synthesis
- Synthesis
- Retrosynthesis
- Cascade reactions

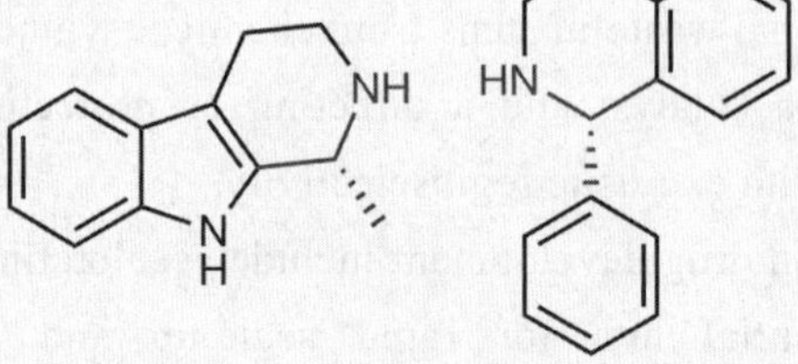

Fig. 18-1 Developments in biocatalysis research that are having an impact on the synthesis of pharmaceutical compounds. Key advances are being made in several areas of enzyme discovery and enzyme engineering, which are leading to an increase in the number and types of enzymes that are available to chemists, thereby providing additional reactions in organic synthesis. Biocatalysts are now being used in both discovery chemistry and process chemistry, including in flow format using immobilized enzymes.

1. Biocatalysis in drug development

In the pharmaceutical industry, various key drivers affect the discovery and development of new medicines and their manufacture in a cost-effective and safe manner. One of these drivers is how quickly new drug candidates can be moved through the preclinical and clinical development phases. At these stages of the drug development process, chemistries that enable the preparation of a sufficient quantity (0.1~1 kg) of a candidate drug for further studies must be quickly identified. Many of these drug candidates are stereochemically complex and have multiple functionalities, and indeed, some candidates belong to classes of molecules termed modalities, which are not typical small-molecule active pharmaceutical (APIs).

Biocatalysis has primarily been used to enable the scalable, cost-effective and controlled synthesis of complex single-stereoisomer drugs. However, this approach to synthesis is changing, enabled by the true power of biocatalysts—their exquisite selectivity in catalysing reactions, which not only avoids the need for protecting group manipulation but also allows for multiple reactions to be run in one pot. Running reactions in tandem rapidly builds molecular complexity and drives down costs, thereby delivering medicines to patients much sooner. Additional newly identified biocatalytic enzymes, such as reductive aminases and a broader range of aldolases, enable the stereoselective formation of molecules, and biocatalytic cascades can build entire pharmaceutical molecules from simple building blocks in a single process.

In addition to the synthesis of complex molecules, biocatalysis can accelerate the drug discovery process by generating lead compounds that have new activities and metabolic profiles. This is an area that has not been fully exploited, probably owing to concerns that the restricted substrate scope of some classes of enzymes limits their success in parallel biocatalytic synthesis of large libraries of compounds. Broader-scope and 'molecule-building' enzymes will no doubt have a greater role in future drug discovery efforts.

2. Enzyme discovery

Major developments in the biocatalysis field in the past 10~15 years have resulted in a rapid diversification of the enzymes that are available to the synthetic chemist. For example, the widespread availability of sequenced genomes is one of the most important developments, which has resulted in an explosion in the number of gene and protein sequences that can be used as the basis for discovering new enzymes. Combined with the increasingly lower cost of DNA sequencing and artificial gene synthesis, synthesizing genes is now a rapid, cost-effective way of generating novel biocatalysts for screening. Consequently, the challenge in the field has shifted from simply discovering new open reading frames (ORFs) to curating existing genomic databases to identify or predict the enzyme activity that is encoded by an ORF.

Bioinformatics. To relate the primary amino acid sequence and the structure of biocatalysts to their function, various databases and algorithms that are based on protein structure (for example, 3DM from Bio-Prodict) or on both primary amino acid sequence and protein structure (for example, Catalophor and Zymphore) have been developed. Databases can also be used to identify target residues for mutagenesis, and the CAVER algorithm has been developed to predict the existence of tunnels (that is, clefts, cavities and pockets) in enzymes, which may be important for substrate access or product release. In many cases, the primary amino acid sequence of an enzyme is not sufficient to predict the activity of an enzyme, and therefore, high-throughput screening methods must be used to identify the activity required (see below).

Metagenomics. New enzymes can also be identified by cloning large metagenomic libraries; this approach is particularly effective when there are clear sequence motifs that can be used to identify homologues. A number of companies (for example, Prozomix and c-LEcta) now produce large panels of biocatalysts (>500 enzymes) from specific enzyme families (such as KREDs, transaminases and imine reductases) using this approach.

De novo design. De novo protein design has made possible major advances in generating new enzyme activities from alternative scaffolds. Combining de novo design and directed evolution can lead to the creation and optimization of new biocatalysts for chemical synthesis. For example, a retro-aldolase (RA95.5) with a catalytic efficiency approaching that of natural aldolases has been generated through initial design using the Rosetta program (a software suite for computational modelling and analysis of protein structures), followed by several rounds of high-throughput directed evolution.

Molecular dynamics. During catalysis, many enzymes undergo conformational motion, such as the movement of loops or domains, within the protein, which may be required for substrate binding or for the release of a reaction product or other factors that are essential for catalysis. Using high-end computing, molecular dynamics is increasingly being used to simulate the movement of loops and domains in the enzyme during the catalytic pathway. Methods to deconvolute activity, selectivity and stability data for enzymes are also important for these analyses. Statistical deconvolution is currently the major paradigm in biocatalysis research; however, driven by the vast amounts of robust data that are being generated in these simulations, artificial intelligence and machine learning approaches are gaining traction.

Enzyme promiscuity. Enzyme promiscuity (the ability of an enzyme to catalyse a side reaction in addition to its main reaction) also has potential to advance the discovery of new biocatalysts. Promiscuous activities of an enzyme are usually slow relative to the main activity and are subject to neutral selection; for example, atrazine chlorohydrolase (encoded by atzA) from Pseudomonassp. ADP evolved from melamine deaminase (encoded by triA), which has promiscuous activity towards the synthetic herbicide atrazine.

Ancestral sequence reconstruction. Ancestral enzymes that existed millions of years ago are generally considered to have been more stable and more promiscuous in their activity (although probably at the expense of lower activity) than extant enzymes. In a method termed ancestral sequence reconstruction (ASR), ancestral sequences can be recreated from the sequence of extant enzymes by making assumptions about mutational drift. These reconstructed ancestral enzymes are likely to be more highly evolvable, as was demonstrated for hydroxynitrile lyases and transaminases.

3. Enzyme engineering

Directed evolution. Directed evolution is a key enabling technology that underpins the development of new biocatalysts for the pharmaceutical and related industries. Existing approaches for high-throughput screening of biocatalyst libraries are based on using enzyme-expressing bacterial cells arrayed in microplates and measuring reactions using high-performance liquid chromatography (HPLC), HPLC mass spectrometry (HPLC MS) or optical detectors, which require either long cycle times or tailor-made substrates. In some cases, bacterial colonies can be screened directly on mem-branes, provided that a colorimetric or fluorescence-based readout can be used to monitor the required activity. The generation of libraries of enzyme variants is now routine and is largely carried out using PCR-based methods. Usually, the full complement of 20 naturally occurring amino acids are used, although this leads to very large libraries of enzyme variants ($20n$; n = the number of amino acids in the enzyme). Library size can be reduced by restricting codon usage (for example, by utilizing the NDT codon degeneracy (D = A, T or G)), so that only a subset of the 20 natural amino acids (for example, 12 amino acids) are encoded. This approach markedly reduces the number of enzyme variants that must be screened, enabling more rapid identification of hot spots (that is, regions that produce substantial changes in activity or selectivity when mutated) within the enzyme. In a study demonstrating the benefits of directed evolution, multiple rounds of mutagenesis of an (S)- transaminase were carried out to identify variants that could accommodate sterically bulky substrates. The resulting enzymes had up to 8,900-fold higher activity than the starting scaffold and were highly stereoselective (up to 99.9% enantiomeric excess) in the asymmetric synthesis of a set of chiral amines containing bulky substituents.

High-throughput screening. The use of microfluidics and microdroplets can lead to a step-change in the number of enzyme variants (up to 108) that can be screened for activity. In these methods, bacterial cells expressing an enzyme are encapsulated together with their substrate in microdroplets and are then screened for catalytic activity using fluorescence-activated cell sorting (FACS)-based methods. Droplet microfluidics interfaced with electrospray ionization (ESI)–MS provides a label-free high-throughput screening platform; for example, optimization of this method by reducing carry-over between droplets (by manipulating the type of injection needle and the sheath flow), which is traditionally an important limitation in these methods, enabled the efficient and successful high-throughput screening of two libraries of transaminases expressed in bacteria, as well as a library of in vitro transcribed and translated transaminases.

Furthermore, desorption ESI–MSI (DESI–MSI) is an innovative ion mobility MS imaging application that was used to screen biotransformation reactions on membranes in real time, for example, for detecting the activity of phenylalanine ammonia lyases (PALs) and cytochrome P450 monooxygenases. A major challenge in enzyme evolution is navigating sequence-activity space in a way that allows the trajectory of fitness (for example, catalytic efficiency and selectivity) to be predicted. Each round of mutagenesis, screening and amplification generates large amounts of data that must be fitted to a machine learning algorithm to facilitate identification of the next sequence that should be constructed; for example, CodeEvolver (Codexis) was developed for this purpose and was used to evolve many biocatalysts that have improved attributes, such as increased solvent tolerance and catalytic activity.

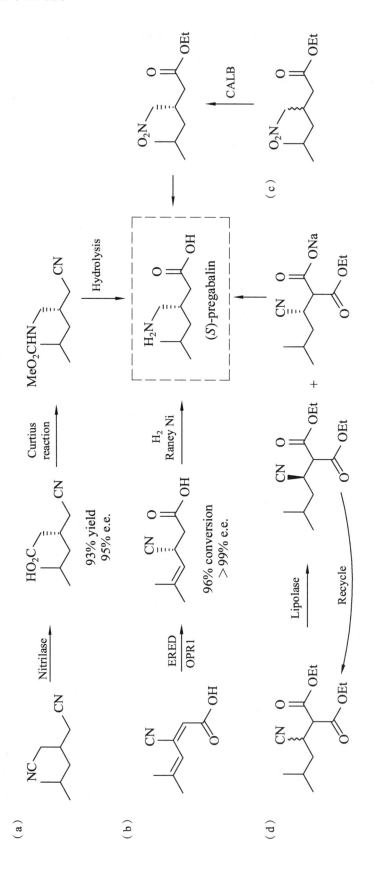

Fig. 18-2 Biocatalytic routes to the synthesis of (S)-pregabalin. (S)-pregabalin (Lyrica; Pfizer) is a medication for the treatment of epilepsy, neuropathic pain, fibromyalgia and generalized anxiety disorder. Several biocatalytic routes, based on the use of both hydrolytic and redox enzymes, have been developed for the synthesis of this active pharmaceutical ingredient (API). a. Hydrolysis-rearrangement-hydrolysis. b. Stepwise exhaustive hydrogenation. c. Enantiomeric enrichment. d. Epimerization followed by reduction. CALB, Candida antarctica lipase B; e.e., enantiomeric excess; ERED, ene reductase; OPR1, 12-oxophytodienoate reductase 1.

4. Biocatalytic retrosynthesis

Retrosynthesis is a commonly used strategy in organic synthesis for planning efficient synthetic routes to target molecules. Retrosynthesis involves disconnection of target molecules, followed by the use of known chemical transformations to optimally reconstruct a molecule from simpler building blocks. The availability of a wide range of reagents for synthetic transformations in organic synthesis means that retrosynthetic analysis can often lead to several different options or pathways, which requires that choices are made about which route is most efficient. Advances in enzyme discovery and protein engineering have enabled the identification of additional retrosynthetic pathways that use biocatalysts in the key bond-forming steps of synthesis after molecules are disconnected. Owing to the complementary nature of chemocatalysts and biocatalysts in terms of reaction chemistry, biocatalytic retrosynthesis can often lead to completely new disconnections, and thus starting materials, compared with conventional organic retrosynthesis. Often, multiple synthetic routes towards a given target molecule can be developed, each involving the use of a biocatalyst in the key step; for example, several different biocatalytic routes for the synthesis of the anti-epileptic and anti-anxiety drug pregabalin (Lyrica; Pfizer) have been reported, including those based on a nitrilase (Fig.18-2a), an ene reductase (Fig.18-2b), a lipase (Fig.18-2c,d) and a transaminase.

To fully embrace the power of biocatalysis, a cultural change among synthetic organic chemists is necessary. Both in academic research and in the pharmaceutical industry, cross training in synthetic organic chemistry and biocatalysis is needed until the point is reached at which biocatalysis and chemocatalysis together are simply considered to be catalysis. If every synthetic chemist considers chemocatalytic and biocatalytic approaches in their retrosynthetic analysis, many more opportunities for increasing the number of routes to a desired product will be identified.

5. New therapeutic modalities

The pharmaceutical industry, biotechnology companies and academic researchers are investigating novel therapeutics to meet the challenge of treating intractable biological targets. These intractable targets include molecules without well-defined active sites or binding pockets that are suitable for the binding of small molecules, molecules with large flat surface interactions and complex intracellular molecules that are inaccessible to antibodies (which can not traverse the plasma membrane) and that lack a well-defined pocket for the binding of small molecules. In addition, the pharmaceutical industry is investigating new modalities that bridge the gap between the two traditional classes of pharmaceutical compounds, namely, small molecules and biologics. These new modalities are typically larger (5-100 kD) than existing small molecules (<0.9 kD) but are smaller than biologics (>150 kD), and they include antibody-drug conjugates (ADCs; comprising a monoclonal antibody, a warhead and a linker), synthetically modified monoclonal antibodies, conjugations and ligations, antisense oli-gonucleotides (ASOs), cyclic dinucleotides

(CDNs) and cyclic peptides. Furthermore, these new modalities encompass drug candidates that are stereochemically and structurally complex, which makes them costly to manufacture at present. Consequently, substantial commercial interest exists in finding novel ways to synthesize these complex molecules. Owing to the complexity of these new modalities, enzymes might be more appropriate catalysts for the synthesis of these modalities than chemocatalysts, and indeed, early proof-of-concept studies suggest that synthetic oligonucleo-tides that contain unnatural bases can be prepared on a small scale using DNA polymerases. Enzymes have also been used to couple peptide fragments to make synthetic oligopeptides and to conjugate small molecules to produce larger biomolecules for targeting drugs to receptors and tumours. Furthermore, biocatalytic technologies have additional benefits for the manufacture of APIs and biomolecules, including reduced waste streams, energy costs and costs of goods, including solvents.

6. New biocatalytic reactions

The continuing usefulness of biocatalysis in organic synthesis depends on the discovery of new reaction chemistries for target molecule synthesis. Historically, biocatalytic enzymes were limited primarily to hydro-lytic enzymes (for example, lipases, esterases and proteases) and redox enzymes (for example, alcohol dehydrogenases). However, in the past 10 years, a range of new enzyme families have been developed, including transaminases, nitrilases, ene reductases, amine oxi-dases, Baeyer-Villiger monooxygenases and aldolases. The rate at which new biocatalytic reactions are being discovered or developed through protein engineering and/or directed evolution continues to increase. In the past 5 years, new biocatalysts have been reported for a wide range of reactions, including C–H amination, reductive amination, carbon-silicon bond formation, insertion of carbenes in C—H bonds, enantioselective halogenation and dehalogenation, trifluoromethylation and many others.

c Olefination of aldehydes

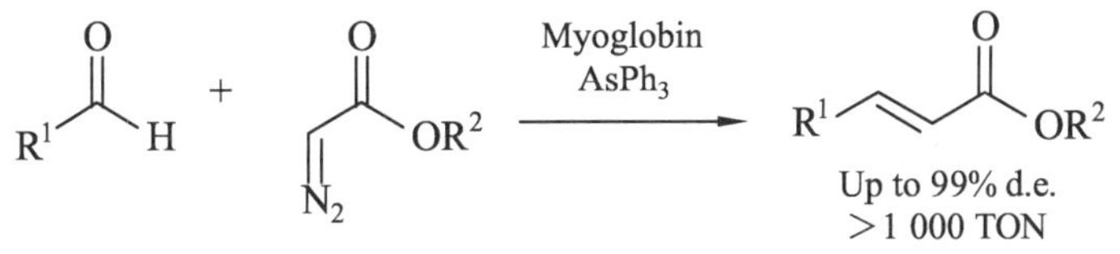

$$R^1CHO + N_2=CHCO_2R^2 \xrightarrow{\text{Myoglobin, AsPh}_3} R^1CH=CHCO_2R^2$$

Up to 99% d.e.
>1 000 TON

d Insertion into C—H bonds via nitrenes

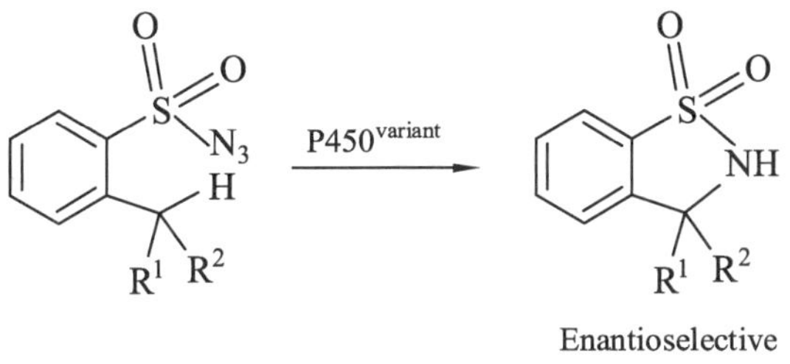

Reaction of 2-(CR^1R^2H)-benzenesulfonyl azide with P450^{variant} gives the cyclized benzisothiazole 1,1-dioxide (sultam) bearing R^1, R^2 at the sp^3 carbon.

Enantioselective

e C—Si bond formation

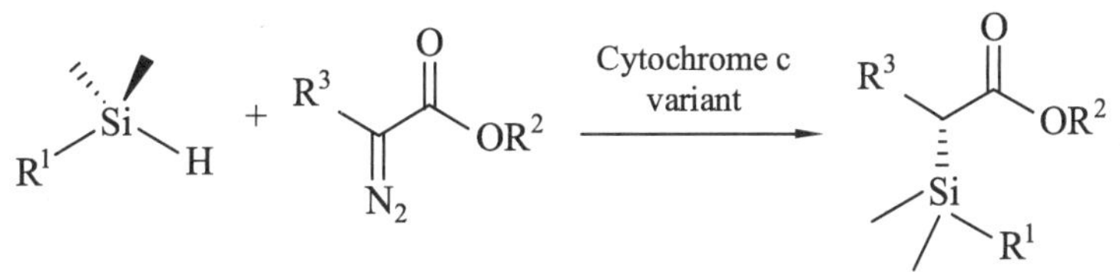

$$R^1R(Me)_2Si-H + N_2=CR^3CO_2R^2 \xrightarrow{\text{Cytochrome c variant}} R^3CH(SiR^1Me_2)CO_2R^2$$

f C—B bond formation

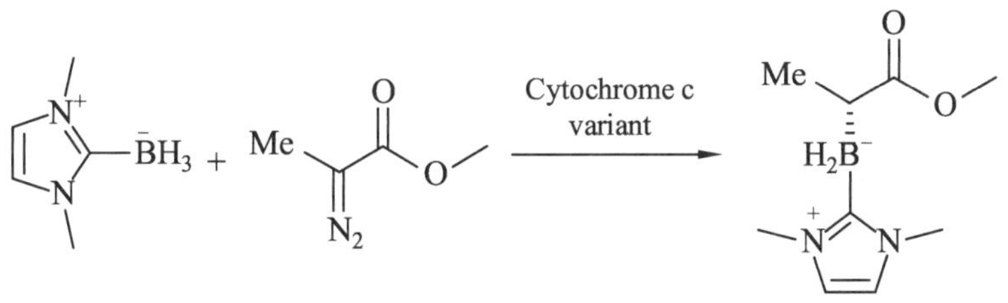

N-heterocyclic carbene borane (1,3-dimethylimidazol-2-ylidene·BH$_3$) + Me(N$_2$)CCO$_2$Me → (1,3-dimethylimidazolium-2-yl)BH$_2$—CH(Me)CO$_2$Me, with Cytochrome c variant.

g Imine reduction

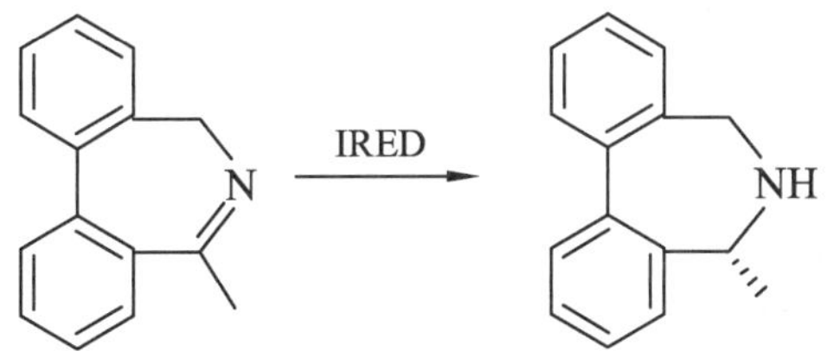

Dibenzazepine-type cyclic imine (with methyl at C=N) → corresponding chiral amine via IRED.

h Reductive amination

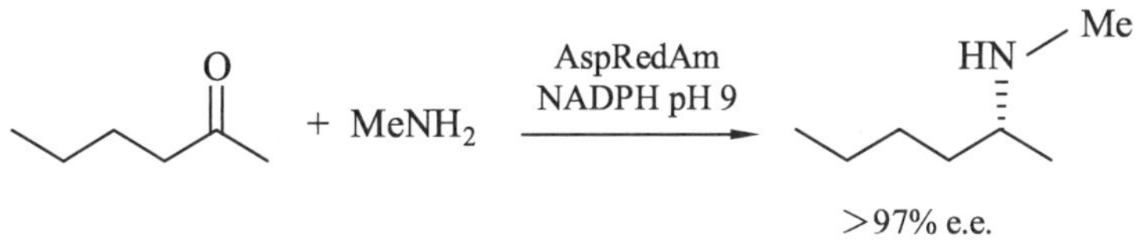

$$\text{2-hexanone} + \text{MeNH}_2 \xrightarrow[\text{NADPH, pH 9}]{\text{AspRedAm}} \text{(R)-N-methyl-2-hexylamine}$$

>97% e.e.

i Alkylation

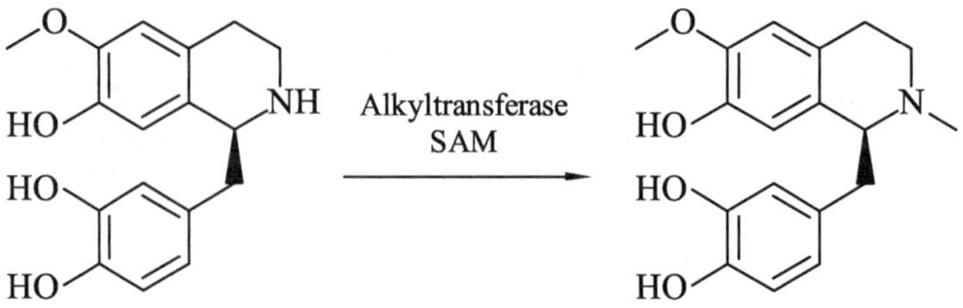

j Pictet-Spengler cyclization

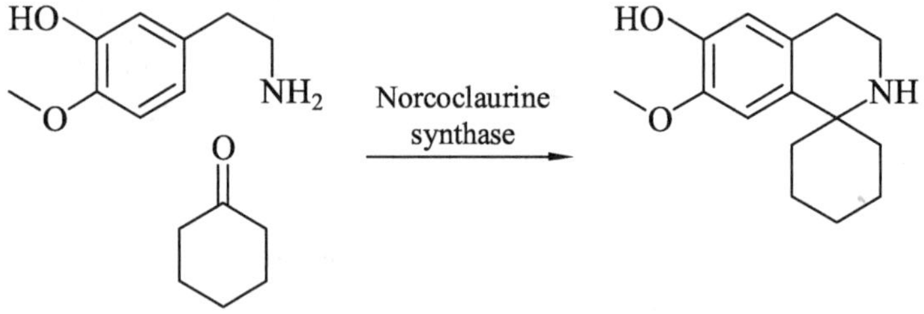

Fig. 18-3 New reactions enabled by biocatalysis: transfers of carbene and nitrene intermediates and synthesis of amines. Examples of new biocatalysts that were either discovered or engineered to catalyse reactions that have previously been carried out using chemocatalysts only, thereby increasing the number of opportunities to use "biocatalytic retrosynthesis". a–c. Transfer of carbene intermediates. d. Transfer of nitrene intermediates. e, f. Carbon–heteroatom bond formation. g–j. Synthesis and derivatization of amines. Of note, some of these reactions are not currently known to occur naturally (parts a–f). AspRedAm, Aspergillus sp. reductive aminase; d.e., diastereomeric excess; e.e., enantiomeric excess; GDH, glucose dehydrogenase; IRED, imine reductase; P450variant, cytochrome P450variant; SAM, S-adenosyl methionine; TON, turnover number.

These newly discovered biocatalytic enzymes (Figs.18-3, 18-4, 18-5) are being increasingly used in discovery chemistry in addition to process chemistry. For biocatalysts to be usefully applied in discovery chemistry, they must have broad substrate specificity to enable the processing of multiple substrates, thereby providing chemists with access to new chemical space. Historically, enzymes have been viewed as having high selectivity (for example, stereoselectivity, chemoselectivity and regio selectivity) but low substrate promiscuity. However, directed evolution and protein engineering have been used together to broaden the substrate scope of existing enzymes by carrying out multiple rounds of library generation and screening against a broad set of substrates. In directed evolution, the substrate of interest can be changed, a process that is often termed substrate walking, to enable the identification of enzyme variants that have a complementary substrate scope.

a Carboxylation

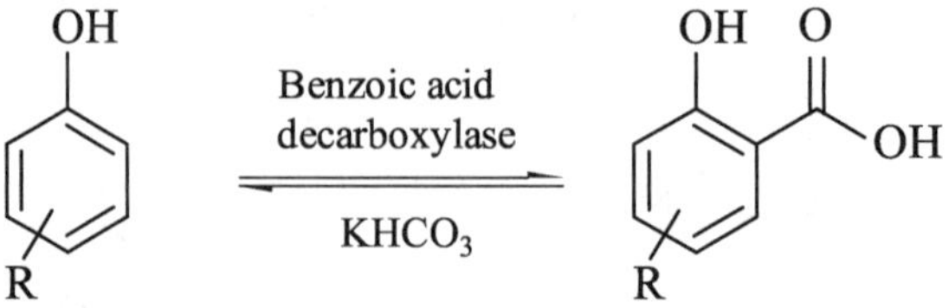

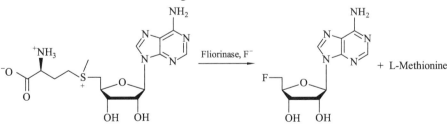

Fig. 18-4 New reaction chemistries enabled by biocatalysis: synthesis and conversions of carboxyl and organic halide functionalities. Examples of new biocatalysts that were either discovered or engineered to catalyse reactions that have previously been carried out using chemocatalysts only, thereby increasing the number of opportunities to use biocatalytic retrosynthesis. a–d. Synthesis and conversion of carboxyl functionalities. e–g. Synthesis and conversion of organic halide functionalities. CAR, carboxylic acid reductase; e.e., enantiomeric excess; KRED, ketoreductase.

7. Expanding biocatalytic chemistries

Cytochrome P450 monooxygenases. Although historically they have mostly been used to catalyse reactions such as hydroxylation, epoxidation, dealkylation and sulfoxidation, cytochrome monooxygenases have proved to be extraordinarily versatile biocatalysts that can be engineered to catalyse a much wider range of reactions. For example, an active-site variant of cytochrome BM3, in which Cys is replaced by His, can use suitable precursors (for example, ethyldiazo-acetate) to generate carbenes, which can then react with suitable alkenes to generate enantiomerically pure cyclo-propanes (Fig.18-3a). In addition, myoglobin-derived cyclopropanation biocatalysts have been used for the highly enantioselective synthesis of various cyclopropane-containing drugs, and replacing the haem domain of both a cytochrome monooxy genase and myo-globin with a transition metal centre (Me–Ir) produced highly active biocatalysts for C—H insertion reactions (Fig.18-3b). Variants of myoglobin were also used to catalyse the formation of C=C bonds between diazo esters and aldehydes (Fig.18-3c) . Variants of cytochrome have been used to convert sulfonyl azides into nitrenes, reactive intermediates that then undergo enantioselective intermolecular and intramolecular C—H amination reactions (Fig.18-3d) .

a   Carbocation cyclization

b   Trifluoromethylation

c   Friedel–Crafts acylation and Fries reaction

d   Hydration of styrenes

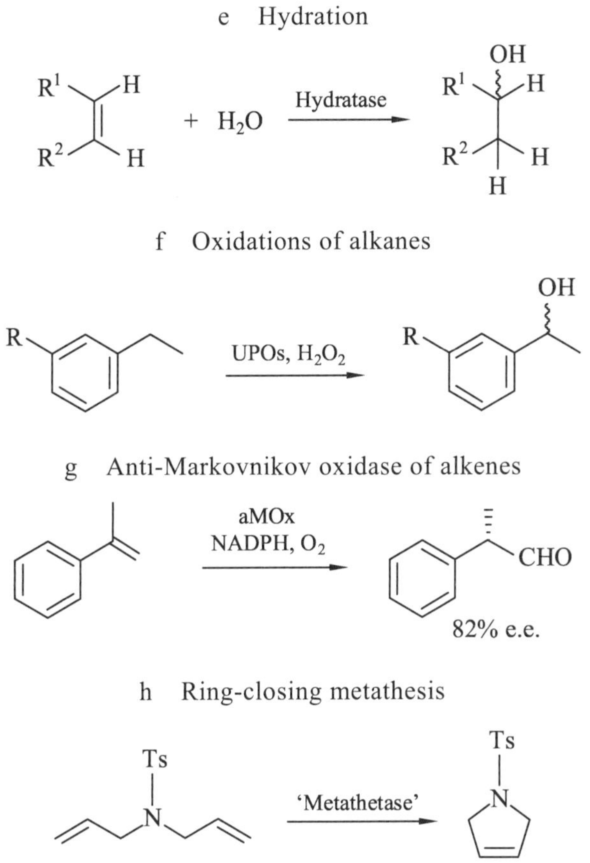

e Hydration

$$R^1R^2C=CH_2 + H_2O \xrightarrow{\text{Hydratase}} R^1R^2C(OH)-CH_3$$

f Oxidations of alkanes

(R-C₆H₄-CH₂CH₃) →[UPOs, H₂O₂]→ (R-C₆H₄-CH(OH)CH₃)

g Anti-Markovnikov oxidase of alkenes

(PhC(CH₃)=CH₂) →[aMOx, NADPH, O₂]→ (PhCH(CH₃)-CHO) 82% e.e.

h Ring-closing metathesis

Ts-N(CH₂CH=CH₂)₂ →['Metathetase']→ N-Ts pyrroline

Fig. 18-5 New reaction chemistries enabled by biocatalysis: miscellaneous chemistry. Examples of new biocatalysts that were either discovered or engineered to catalyse reactions that have previously been carried out using chemocatalysts only, thereby increasing the number of opportunities to use biocatalytic retrosynthesis. a. Carbocation cyclization. b. Trifluoromethylation. c. Friedel–Crafts acylation and Fries reaction. d, e. Alkene hydration reactions. f, g. Oxidations of alkanes and alkenes. h. Ring-closing metathesis. aMOx, anti- Markovnikov oxidase; e.e., enantiomeric excess; PAD, phenolic acid decarboxylase; SHC, squalene–hopene cyclase; TFMS, trifluoromethyl silane; Ts, 4-toluenesulfonyl; UPOs, unspecific peroxygenases.

Cytochrome enzyme variants can also catalyse other C-H aminations (for example, those involving amides), as well as C–Si (Fig.18-3e) and C—B bond formation (Fig.18-3f) . Reductions. Imine reductases (IREDs) were first described in 2010, and since then, there has been tremendous interest in the potential use of these enzymes for asymmetric imine reduction (Fig.18-3g) . The scope of this class of enzyme has been expanded to include reductive amination (catalysed by reductive aminases (RedAms)) that involves the coupling of selected ketones and amines at 1:1 stoichiometry to generate enantiomerically enriched amine products (Fig.18-3h) . Alkyltransferases can be used to decorate the amines that are produced to generate more complex alkaloids (Fig.18-3i) , which can themselves be derived from asymmetric biocatalytic Pictet–Spengler reactions (Fig.18-3j) . Reactions of carboxy acids and esters. Biocatalysts are being developed for both enzyme-mediated carboxylation (Fig.18-4a) and decarboxylation (Fig.18-4b) . Carboxylic acid reductases (CARs) are modular enzymes that convert carboxylic acids into aldehydes without over-reduction to the corresponding alcohols, in a reaction that is dependent on ATP and NADPH (Fig.18-4c). The published structure of various bacterial CARs will undoubtedly aid further protein engineering of these enzymes. Furthermore, CARs can be engineered for amide synthesis activity, suggesting a new application of these enzymes (Fig.18-4d). Indeed, the frequency of amides in APIs has made the discovery and development of biocatalysts for amide synthesis a high priority for the pharmaceutical industry.

Halogenation and dehalogenation. Halogenases typically require electron-rich aromatic and hetero-aromaticsubstrates (for example, pyrroles, indoles and phenols) and use both chloride and bromide ions as the halogen source (Fig.18-4e). A key breakthrough in halogenase biocatalysis was made by the co-immobilization of the requisite recycling enzymes, thereby enabling gram-scale synthesis. Furthermore, engineered halogenase enzymes can be integrated with palladium catalysis to enable one-pot halogenation and C—C bond-forming reactions. Enantioselective halogenation for desymmetrization of a prochiral substrate and enantioselective light-driven dehalogenation of race-mic halolactones using KREDs (Fig.18-4f) have also been reported. Finally, although fluorinases are somewhat limited in substrate scope, fluorination is a highly challenging and important reaction, and further development of these enzymes could enable their wider application in synthesis (Fig.18-4g).

Other reactions. Terpene cyclases are a very broad family of enzymes that are responsible for the conversion of precursors such as isopentenyl pyrophosphate (IPP), geranyl pyrophosphate (GPP) and farnesyl pyro-phosphate (FPP) to the thousands of different naturally occurring terpene-based compounds. Terpene cyclases have been shown to accept nucleophiles (for example, OH and enols) other than C=C bonds in the cationic cyclization process, thereby providing access to novel products (Fig.18-5a). These enzymes have also been engineered to alter the product distribution of specific terpene products. For example, two new reactions that use biocatalysis have been developed, namely, trifluoromethylation (Fig.18-5b) and an intriguing Friedel-Crafts-type acylation of phenols (Fig.18-5c). Hydration of alkenes (Fig.18-5d,e) and dehydration of alcohols, particularly at unactivated positions, offer the potential for functionalization of a broad range of substrates, including fatty acids and alkanes. The published structure of a ligand-bound bacterial bifunctional linalool dehydratase isomerase has provided insights into the catalytic mechanism of these enzymes. Unspecific peroxygenases (UPOs) are a family of haem-thiolate enzymes that have similar reaction chemistry to cytochrome monooxygenases but use hydrogen peroxide instead of molecular oxygen as the oxidant (Fig.18-5f). UPOs are typically found in fungi (for example, Agrocybe sp.) and hence must be expressed in fungi such as Pichia pastoris, Saccharomyces cerevisiae or Aspergillus nidulans to obtain a functional enzyme. UPOs can be combined with photocatalytic water oxidation for the generation of hydrogen peroxide. Engineered cytochrome enzymes have been shown to catalyse anti-Markovnikov oxidation of styrenes to the corresponding ketones, a reaction that is difficult to recapitulate using conventional chemical catalysis (Fig.18-5g) . Artificial metalloenzymes are beginning to be used in reactions for which natural or even evolved enzymes are not currently available. The first example of a metathesis reaction in vivo using the streptavidin concept with a Ru catalyst, as well as the first combination of an artificial metalloenzyme with various enzymes in a concurrent-type cascade (Fig.18-5h) , has been reported.

In parallel to the creation of new biocatalytic activities, there is considerable interest in finding new applications for existing enzymes, including the development of "smart substrates" that improve the efficiency of existing transformations. For example, transaminase reactions are notoriously limited by poor reaction equilibrium, resulting in the starting material often being favoured over the product. The use of specific amine donors can lead to enhanced conversions by providing a driving force for conversion of the ketone substrate into an amine product. Similar smart substrates have been used in KRED-mediated processes.

Lead diversification and library synthesis. To meet the accelerating demands of pharmaceutical discovery, medicinal chemists usually functionalize a core template at predefined positions to explore structure-activity relationships and to fine-tune the desired pharmaceutical attributes. However, without redesigning the template and the synthetic route, using this approach alone restricts the chemical space available and thereby potentially delays the identification of a lead. Advances in biocatalysis now enable late-stage functionalization in a single step, providing quick assessment of a diverse chemical space with the consumption of only milligrams of material (Fig.18-6). Examples of late-stage functionalization include microbial hydroxylation and demethylation to generate phosphodiesterase 2A inhibitors with reduced drug-drug interaction. Similarly, a combination of cytochrome catalysed hydroxylation and chemical halogenation enabled the late-stage replacement of hydrogen with metabolically stable fluorine in various drugs, which blocked drug metabolism and enabled further probing of chemical space.

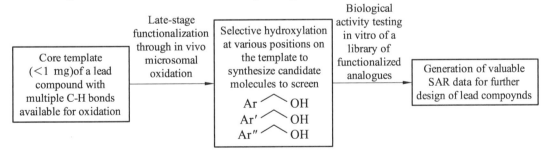

Fig. 18-6 Lead diversification through late-stage functionalization using biocatalysts. Biocatalysts, in particular cytochrome monooxygenases, are being increasingly used in late-stage functionalization of lead compounds through C-H activation in vivo in microsomes, thereby enabling new structure–activity relationship (SAR) data to be obtained. Ar, aryl.

Combinatorial chemistry. Combinatorial chemistry was an area of great interest at the turn of the 21st century and often involved the synthesis of libraries of biologically active compounds in a pooled one-pot reaction. Biocatalysis proved useful for expanding the diversity of these chemically derived libraries by facilitating a multitude of lead diversification biotransformations. However, burdensome synthetic and analytical challenges in the identification of hits have resulted in these techniques being surpassed in the past 10–15 years by better annotated and far larger compound libraries, particularly those encoded with strands of DNA. In these biocatalytic libraries, molecular biology techniques are not only used for the DNA encoding itself, but on- DNA chemistry is particularly well suited for the mild and aqueous conditions that are typically used in biocatalysis. For example, oligosaccharides were assembled on DNA using glycosyltransferases, thereby enabling the rapid, simple generation of annotated libraries of carbohydrates.

Parallel medicinal chemistry. Biocatalysis is underexploited in PMC, despite many classes of enzyme showing broad substrate specificity (for example, KREDs). However, a screen of ~150 amines was carried out by an academia-industry collaboration using three established variants of monoamine oxidase N (MAO-N) termed D5, D9 and D11. These three variants were developed to oxidize primary, secondary and tertiary amines to the corresponding imines or iminium ions. Interestingly, the hit rate observed was ~45%, and of these hits, a total of 20 reactions was carried out on apreparative scale.

8. Cascade reactions

The availability of a much broader range of biocatalysts has raised the possibility of building entirely new enzymatic pathways, both in vitro and in vivo, for the conversion of simple, inexpensive starting materials to more complex target molecules, including APIs. These cascade processes can be conceived and designed through a biocatalytic retrosynthetic approach in which the target compound D is disconnected into its constituent parts A, B and C, enabling the identification and engineering of specific biocatalysts for application in the cascade (Fig.18-7). Construction of the cascade followed by cycles of optimization results in a multi-enzyme process for the synthesis of complex molecules. These synthetic cascades constructed in the laboratory are clearly distinct from natural biosynthetic pathways that have evolved for the production of natural products, including morphine alkaloids. A number of important features must be incorporated into the design of biocatalytic cascades: the overall thermodynamic parameters of the cascade must be favourable (ΔG cascade < 0); the enzymes that are selected for the cascade must catalyse the transformations with high reaction specificity and functional group orthogonality to avoid unwanted cross reactivity between different substrates; and the over-all kinetic reaction parameters must be controlled to ensure efficient flux of a substrate or substrates to form a product. In an example of a designed cascade that has been used to generate a range of different products, a combination of CAR, transaminase and IRED biocatalysts were used in the synthesis of substituted chiral piperidines (Fig.18-8).

Synthetic cascades have also been generated by combining engineered biocatalysts with artificial transfer hydrogenases, as well as supramolecular catalysts, although compartmentalization of the individual catalysts within a host (for example, a biotinylated chemo-catalyst compartmentalized

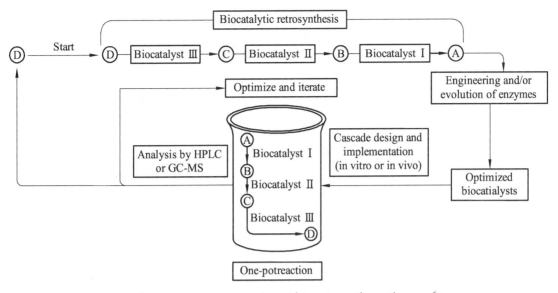

Fig. 18-7 Cascade reactions incorporating biocatalytic retrosynthesis. The use of two or more biocatalysts in a cascade process enables the conversion of relatively simple starting materials into more complex intermediates for the synthesis of active pharmaceutical ingredients (APIs). Biocatalytic retrosynthesis leads to the identification of synthons (that is, building blocks) that inform the selection of engineered enzymes that can catalyse the forward synthetic transformation. These enzymes can then be assembled in a cascade, either in vitro in a one-pot reaction or in vivo. GC-MS, gas chromatography-mass spectrometry ; HPLC, high-performance liquid chromatography.

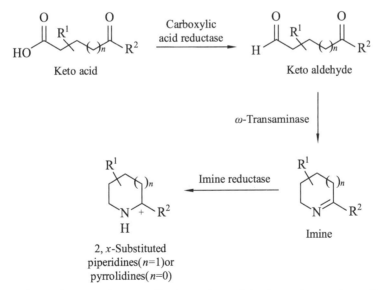

Fig. 18-8 A biocatalytic cascade process for the synthesis of chiral piperidines and pyrrolidines. An example of a synthetic cascade process in which three different enzymes are used in sequence to convert simple achiral starting materials into more complex chiral substituted piperidines or pyrrolidines is depicted. First, a carboxylic acid reductase converts the keto acid into the keto aldehyde, followed by transamination of the aldehyde group into the corresponding amine using an ω-transaminase. The amino ketone then spontaneously cyclizes into the imine, which undergoes asymmetric reduction into the cyclic amine in the presence of an imine reductase. R 1 = R 2 = Me.

within streptavidin) was key to a successful outcome. In an approach that was inspired by the "hydrogen-borrowing" concept in organic chemistry, a cascade has been developed for the redox-neutral alkylation of amines using alcohols.

9. Biocatalysis in flow format

With the expanded opportunities for using biocatalysis for asymmetric synthesis, a number of complementary technologies have come to the fore for the development of intensified and industrially relevant biocatalytic processes. In particular, there is developing interest in carrying out enzyme-catalysed transformations in continuous flow (Fig.18-9) .The application of flow chemistry relies on the continuous pumping of reagents through a reactor zone, which is an area that contains the catalysts. Compared with batch processes, transformations carried out in continuous flow have improved reaction productivity, efficiency of scale-up and downstream processing. Furthermore, continuous flow processes offer the following potential benefits: excellent control of temperature, pressure and mixing; in situ product separation and/or recycling of unreacted reagents; real-time online optimization of reaction conditions to maximize efficiency; real-time in-line monitoring that enables the detection and correction of process condition deviations, resulting in less of the batch being at risk of being discarded; and integration with complementary enabling techno logies such as microwave, photochemical and supported catalyst reactions.

For biocatalytic transformations in particular, continuous flow can reduce enzyme inhibition, facilitate downstream processing when using immobilized bio-catalysts (no lixiviation), improve total turnover numbers (TONs) and enhance volumetric productivity through the ability to run a reaction at the steep part of the Michaelis–Menten kinetic curve.

Preliminary reports describing the application of flow chemistry in biocatalysis involved the use of well-characterized enzyme classes, such as lipases, KREDs and transaminases. Cofactor-dependent enzymes, such as amine dehydrogenases, have also been used in flow chemistry by immobilizing the enzyme on controlled-pore glass using the EziG system from EnginZyme. Clearly, many unmet challenges in this area of biocatalysis exist, including cofactor recycling, mass transfer (solid–liquid–gas in some cases), engineering enzymes for immobilization (are better supports needed?) and matching reaction rates during the process to avoid the use of holding tanks.

Biocatalytic retrosynthesis and enzyme selection

Input
Target compound:
· In silico analysis of pathways and databases
· Pruning of data
· Ranking, scoring and confidence interval

Output
Continuous biocatalytic route

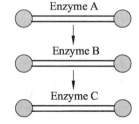

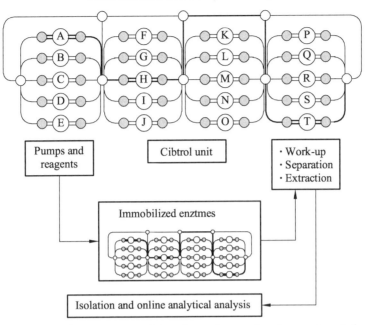

Fig. 18-9 multistep synthesis using immobilized biocatalysts in flow format. A work flow for using immobilized engineered biocatalysts in flow format to enable the establishment of cascade reactions is shown. Biocatalytic cascade reactions (see Fig.18-7) carried out in flow format enable greater control over the individual enzyme-catalysed steps than reactions carried out in batch format.

10. Future developments

In this Review, we highlighted recent developments in the field of biocatalysis that, in our opinion, are driving the increased uptake of this technology within the pharmaceutical industry. As discussed above, there is an increasing need for greater speed and agility when developing new medicines, particularly in areas in which current synthetic approaches are inefficient or too costly.

It is also important to be able to identify new emerging synthetic methods that can be rapidly scaled up and applied for the production of compounds for preclinical studies. Biocatalysis is increasingly viewed as a technology that can contribute to both the discovery of new pharmaceuticals and their preparation on a large scale. A continued increase in the number of biocatalysts and the use of multi-enzyme cascades to enable telescoped reactions to be developed in one pot are crucial if the adoption of biocatalysis as an enabling technology for discovery and process development is to be maintained.

As the cost of gene synthesis continues to fall, we are rapidly approaching a turning point at which it will be more cost-effective to synthesize genes and libraries of enzymes than to use conventional cloning techniques. This turning point is expected to have a major effect on biocatalysis; for example, large panels (~1,000) of biocatalysts can now be created in a matter of weeks, and each variant can be produced on a gram scale for screening. Despite major advances in the high-throughput screening of biocatalysts, this step is still clearly a bottleneck in rapidly assessing the activity of large numbers of enzyme variants. Another challenge is how quickly enzymes can be evolved to a point at which their activity and selectivity are sufficient for preparative-scale synthesis of target compounds. Directed evolution typically takes several months, and the goal is to reduce this step to several weeks by increasingly guiding the process using computational modelling and/or design, sequence-based bioinformatics and machine learning algorithms to rapidly optimize sequences with the desired function. Enzyme production methods are also changing as in vitro translation is being increasingly adopted for the production of enzyme variants. In vitro translation can typically be carried out in 24 hours and thus its use substantially reduces the amount of time needed to evaluate novel biocatalysts. The application of retrosynthetic strategies for planning chemical syntheses, including those using biocatalysts, will undoubtedly increase in sophistication by the use of emerging neural network applications and artificial intelligence tools.

Finally, it is interesting to speculate how a biocatalysis laboratory in a pharmaceutical company, or even an academic laboratory, will differ in 10 years' time. Certainly, there are clear differences in the approaches that are currently used compared with those of 10 years ago—currently, there is a much greater use of robotics for high-throughput enzyme synthesis and screening, and biocatalysis is more highly integrated into the workflow of drug discovery and development. How will biocatalysis develop in the next 10 years? It is conceivable that every chemist, in addition to using an HPLC instrument, will also have an automated biocatalytic synthesizer. This instrument will have two functions that we anticipate will substantially streamline the use of biocatalysts in drug discovery and development. First, a chemist will be able to prepare new enzymes for screening simply by inputting the desired sequence or mutations. We anticipate that the first iteration of this instrument would be capable of benchtop high-throughput DNA synthesis and protein expression for biocatalyst production, which would eliminate delays resulting from the need to ship synthesized DNA constructs and enzymes from vendors, which will probably soon be a rate- limiting step in enzyme engineering (and already is in some cases). Second, the instrument will be able to carry out cascade reactions in flow format, again simply by dialling in the sequence of biocatalysts that is required for the various steps (for example, conversion of compound A into compound D via compounds C and D). The instrument should also have a retrosyn-thesis program that will provide suggestions for different synthesis routes and enzymes that can be used to prepare a target molecule specified by the user. Automation has revolutionized DNA synthesis and sequencing and peptide synthesis and has even been implemented in oligosaccharide synthesis . These developments suggest that automation will have the same effect on the efficiency of biocatalysis in drug development.

Part 5

Supplementary Reading

Appendix 1: Introduction about the Undergraduate Programme in Queen's University

Overview

Chemistry is a core science subject that touches almost every aspect of our daily lives, and will become increasingly important in our future knowledge-based society. Chemists develop life-saving drugs, polymers, pest-control agents and catalysts that can enhance our quality of life beyond measure.

Four-year MSci degrees are available for high-calibre students with the ability and aspiration to practise Chemistry at the highest levels. BSc students with excellent performance may transfer to the MSci up to the end of Stage 2.

Note: the School has also introduced a new degree in Engineering Chemistry which covers core elements of both chemistry and chemical engineering, and will prepare students for a wide range of careers. It is available in two options—the MSci or the MEng, the award dictated by the modules studied.

Chemistry with Study Abroad Degree highlights

Professional Accreditations

The BSc degrees are recognised by the Royal Society of Chemistry and the Institute of Chemistry in Ireland, and the MSci Chemistry is accredited by The Royal Society of Chemistry.

Course content

Course Structure

Introduction

While providing dedicated subject-specific learning, our Chemistry degrees strongly emphasise opportunities to develop generic problem-solving and reflective-working practices applicable to a range of career paths and patterns of employability.

Many of the elements of the BSc are in common with the MSci programme, and allow students to transfer between the two pathways, subject to meeting the appropriate programme requirements.

All degrees are modular, with six modules each year. All provide a thorough training in the three main subject areas (Inorganic, Organic and Physical Chemistry) through compulsory core modules which offer in-depth study of these three areas.

Stage 1

Students study a common programme with the Chemical Engineers, giving them an understanding of how the two subjects relate to each other and an opportunity to transfer if they decide they are better suited to the other discipline. Key to this is a course structure permitting students to study both introductory Chemistry, and Chemical Engineering, alongside a couple of skills modules equipping students to proceed on either degree programme.

In the second semester students then take three modules covering the main fundamental subject areas; inorganic, organic and physical chemistry.

Stage 1 courses are outlined below:
Introduction to Mathematics for Chemists and Engineers
Skills for Physical Chemistry
Organic Chemistry Level 1
Inorganic Chemistry 1
Physical Theory

Stage 2

Students are required to take six modules of chemistry, designed to extend their knowledge of the traditional subject areas of inorganic, organic and physical chemistry, in addition to introducing aspects of applied chemistry, spectroscopy and theoretical chemistry. Each of the modules contain both practical and coursework components allowing students to develop, practice and demonstrate a wide range of professional skills.

Stage 2 courses are outlined below:
Structural Chemistry
Quantum Theory, Spectroscopy and Bonding
Organic Chemistry 2
Inorganic Chemistry 2
Physical Chemistry 2
Industrial and Green Chemistry

Stage 3
Placement Year

Stage 4

In addition to advancing the three main subject areas of organic, inorganic and physical chemistry, students can also select a number of applied options allowing opportunities to specialise.

Students have the choice of taking either a double-weighted research project directly supervised by a member of staff, or experiencing the full breadth of the subject through taking a series of three extended mini-projects in each of the main subject areas, making up a double-weighted module. Key to both of these options is the acquisition of both subject-specific and generic skills to act as a springboard to a successful career.

Different pathways offer opportunities to specialise. In the later stages there are optional specialist modules and extended practical/project work. The specialist pathways available consist of additional elements which are detailed below:

Chemistry with Study Abroad: students take French or Spanish alongside Chemistry in Stages 1 and 2, then spend a year abroad studying Chemistry in French or Spanish, and then return to Queen's for Stage 3.

BSc Sandwich Degrees: students spend their third year working in industry (subject to the availability of a suitable placement), then return to Queen's for a final year of study.

Medicinal Chemistry: students take modules which include Biochemistry, Genetics and Medicinal Chemistry, and undertake a medicinal or biological project.

Core course is outlined below:

Chemical Research Project

Stage 4 Optional Courses

Advanced Organic Synthesis

Advanced Physical Chemistry

Advanced Inorganic Chemistry

Options in Applied Chemistry

People teaching you

Dr. Gary Sheldrake

Director of Education

Chem & Chemical Engineering

Dr. Sheldrake is a senior lecturer in Chemistry. He specialises in Synthetic and Bioorganic Chemistry.

Contact Teaching Times

Large Group Teaching

7 (hours maximum)

7 hours of lectures or seminars; (similar learning outcomes in the Study Abroad institution should apply to all categories)

Medium Group Teaching

6 (hours maximum)

6 hours of practical classes or workshops each week; laboratory hours will increase as more project work is undertaken at Levels 3-4 (as applicable)

Personal Study

24 (hours maximum)

22–24 hours studying and revising in your own time each week, including some guided study using handouts, online activities, etc.

Small Group Teaching/Personal Tutorial

2 (hours maximum)

2 hours of tutorials (or later, project supervision) each week

Learning and Teaching

At Queen's, we aim to deliver a high quality learning environment that embeds intellectual curiosity, innovation and best practice in learning, teaching and student support to enable student to achieve their full academic potential.

On the MSci in Chemistry with Industry we do this by providing a range of learning experiences which enable our students to engage with subject experts and develop attributes and perspectives that will equip them for life and work in a global society. We make use of innovative technologies and a world class library to enhance their development as independent, lifelong learners.

Examples of the opportunities provided for learning on this course are:

E-Learning technologies

Information associated with lectures and assignments is typically communicated via a Virtual Learning Environment (VLE) called Queen's Online. Opportunities to use IT programmes associated with data manipulation and presentation are embedded in the practicals and the project-based work.

Lectures

Introduce basic information about new topics as a starting point for further self-directed private study/reading. Lectures also provide opportunities to ask questions, gain some feedback and advice on assessments (normally delivered in large groups to all year group peers).

Personal Tutor

Undergraduates are allocated a Personal Tutor during Level 1 and 2 who meets with them on several occasions during the year to support their academic and professional development through the discussion of selected topics.

Practicals

These are essential to the training in this laboratory based subject area. You will have opportunities to develop technical skills and apply theoretical principles to real-life or practical contexts. Most of the core taught modules at Stages 1 and 2 have practical components associated with them, whilst stage 3 has a double weighted practical module (CHM3015). Typically at stage 1 you would be in the lab for two afternoons and in stages 2 to 3 it is two full days a week.

Self-directed study

This is an essential part of life as a Queen's student when important private reading, preparation for seminars/tutorials, writing of laboratory reports can be completed. You are encouraged to undertake private reflection on feedback, and at the later stages undertake independent research using the primary literature to support project work and critically review taught course material.

Seminars/tutorials

Significant amounts of teaching are carried out in small groups (typically 6~10 students). These provide an opportunity for students to engage with academic staff who have specialist knowledge of the topic, to ask questions of them and to assess their own progress and understanding with the support of peers. You should also expect to make presentations and other contributions to these groups as well as using them as a route to providing individual feedback.

Supervised projects

In the final year, you will be expected to carry out a significant piece of research on a topic or practical methodology that you have chosen. You will receive support from a supervisor who will guide you in terms of how to carry out your research. The supervisor and a second academic member of staff will formally meet, interview and review the work at the half way stage, and then provide support in the write up stage, although weekly contact is anticipated in most projects within the School.

Assessment

Details of assessments are associated with this course are outlined below:

The way in which you are assessed will vary according to the learning objectives of each module. Some modules are assessed solely through project work or written assignments. Others are assessed through a combination of coursework and end of semester examinations. Details of how each module is assessed are shown in the Student Handbook which is provided to all students through the VLE, and on the School's own website.

Feedback

As students progress through their course at Queen's they will receive general and specific feedback about their work from a variety of sources including lecturers, module co-ordinators, placement supervisors, personal tutors, advisers of study and peers. University students are expected to engage with reflective practice and to use this approach to improve the quality of their work. Feedback may be provided in a variety of forms including:

Feedback provided via formal written comments and marks relating to work that you, as an individual or as part of a group, have submitted.

Face to face comment. This may include occasions when you make use of the lecturers' advertised "office hours" to help you to address a specific query.

Placement employer comments or references.

Online or emailed comment.

General comments or question and answer opportunities at the end of a lecture, seminar or tutorial.

Pre-submission advice regarding the standards you should aim for and common pitfalls to avoid. In some instances, this may be provided in the form of model answers or exemplars which you can review in your own time.

Feedback and outcomes from practical classes.

Comment and guidance provided by staff from specialist support services such as, Careers, Employability and Skills or the Learning Development Service.

Entry Requirements

Entrance requirements

A level requirements

AAB including Chemistry and a second Science subject + GCSE Mathematics grade C.

Irish leaving certificate requirements

H2H3H3H3H3H3 including Higher Level grade H3 in Chemistry and a second Science subject + if not offered at Higher Level then Ordinary Level grade O4 in Mathematics

Access/Foundation Course

Not considered. Applicants should apply for the BSc Chemistry degree.

Selection Criteria

In addition, to the entrance requirements above, it is essential that you read our guidance notes on "How we choose our students" prior to submitting your UCAS application.

How we choose our students

Applications are dealt with centrally by the Admissions and Access Service rather than by the School of Chemistry and Chemical Engineering. Once your on-line form has been processed by UCAS and forwarded to Queen's, an acknowledgement is normally sent within two weeks of its receipt at the University.

Selection is on the basis of the information provided on your UCAS form, which is considered by the Selector for the MSci degree in Chemistry with Study Abroad along with a member of administrative staff from the Admissions and Access Service. Decisions are made on an ongoing basis and will be notified to you via UCAS.

Applicants for the MSci degree in Chemistry with Study Abroad must normally have, or be able to achieve, a minimum of five GCSE passes at grade C or better (to include English Language and Mathematics), though this profile may change from year to year depending on the demand for places. The Selector also checks that any specific entry requirements in terms of GCSE and/or A-level subjects can be fulfilled.

Offers are normally made on the basis of three A-levels. Two subjects at A-level plus two at AS would also be considered. The offer for repeat candidates may be one grade higher than for first time applicants. Grades may be held from the previous year.

Applicants offering two A-levels and one BTEC Subsidiary Diploma/National Extended Certificate (or equivalent qualification) will also be considered. Offers will be made in terms of performance in individual BTEC units rather than the overall BTEC grade(s) awarded. Please note that a maximum of one BTEC Subsidiary Diploma/National Extended Certificate (or equivalent) will be counted as part of an applicant's portfolio of qualifications. The normal GCSE profile will be expected.

Applicants offering other qualifications, such as the International Baccalaureate or Irish Leaving Certificate, will also be considered. The same GCSE (or equivalent) profile is usually expected of those candidates offering other qualifications.

The information provided in the personal statement section and the academic reference together with predicted grades are noted but, in the case of the MSci degree in Chemistry with Study Abroad, these are not the final deciding factors in whether or not a conditional offer can be made. However, they may be reconsidered in a tie break situation in August.

A-level General Studies and A-level Critical Thinking would not normally be considered as part of a three A-level offer and, although they may be excluded where an applicant is taking four A-level subjects, the grade achieved could be taken into account if necessary in August/September.

Candidates are not normally asked to attend for interview.

If you are made an offer then you may be invited to a School Visit Day, which is usually held in the second semester. This will allow you the opportunity to visit the University and to find out more about the degree programme of your choice and the facilities on offer. It also gives you a flavour of the academic and social life at Queen's.

If you cannot find the information you need here, please contact the University Admissions and Access Service (admissions@qub.ac.uk), giving full details of your qualifications and educational background.

INTERNATIONAL STUDENTS

For information on international qualification equivalents, please check the specific information for your country.

English Language Requirements

An IELTS score of 6.0 with a minimum of 5.5 in each test component or an equivalent acceptable qualification, details of which are available at: http://go.qub.ac.uk/EnglishLanguageReqs

If you need to improve your English language skills before you enter this degree programme, INTO Queen's University Belfast offers a range of English language courses. These intensive and flexible courses are designed to improve your English ability for admission to this degree.

Academic English: an intensive English language and study skills course for successful university study at degree level.

Pre-sessional English: a short intensive academic English course for students starting a degree programme at Queen's University Belfast and who need to improve their English.

INTERNATIONAL STUDENTS - FOUNDATION AND INTERNATIONAL YEAR ONE PROGRAMMES

INTO Queen's offers a range of academic and English language programmes to help prepare international students for undergraduate study at Queen's University. You will learn from experienced teachers in a dedicated international study centre on campus, and will have full access to the University's world-class facilities.

These programmes are designed for international students who do not meet the required academic and English language requirements for direct entry.

Careers

Career Prospects

Introduction

Degree plus award for extra-curricular skills

In addition to your degree programme, at Queen's you can have the opportunity to gain wider life, academic and employability skills. For example, placements, voluntary work, clubs, societies, sports and lots more. So not only do you graduate with a degree recognised from a world leading university, you'll have practical national and international experience plus a wider exposure to life overall. We call this Degree Plus. It's what makes studying at Queen's University Belfast special.

Fees and Funding

Tuition Fees

The tuition fee rates for undergraduate students who first enrol at the University in the academic year 2018-19 have not been agreed. Tuition fees for 2019-20 will be based on 2018-19 levels, normally increased by inflation and these are set out below.

Northern Ireland (NI)	£4,160
England, Scotland or Wales (GB)	£9,250
Other (non-UK) EU	£4,160
International	£19,500

Tuition fee rates are calculated based on a student's tuition fee status and generally increase annually by inflation. How tuition fees are determined is set out in the **Student Finance Framework**.

Additional course costs

All Students

Depending on the programme of study, there may be extra costs which are not covered by tuition fees, which students will need to consider when planning their studies.

Students can borrow books and access online learning resources from any Queen's library.

If students wish to purchase recommended texts, rather than borrow them from the University Library, prices per text can range from £30 to £100. A programme may have up to 6 modules per year, each with a recommended text.

Students should also budget between £30 to £75 per year for photocopying, memory sticks and printing charges.

Students undertaking a period of work placement or study abroad, as either a compulsory or optional part of their programme, should be aware that they will have to fund additional travel and living costs.

If a final year includes a major project or dissertation, there may be costs associated with transport, accommodation and/or materials. The amount will depend on the project chosen. There may also be additional costs for printing and binding.

Students may wish to consider purchasing an electronic device; costs will vary depending on the specification of the model chosen.

There are also additional charges for graduation ceremonies, examination resits and library fines.

Chemistry with Study Abroad costs

Students are required to buy a laboratory coat in year 1 at a cost of £15. Students have the option to hire a locker, at a cost of £5 per student per year. Students have the option to join the Royal Society of Chemistry (RSC) at a cost of £19 per annum. Students who undertake a period of study or work abroad, as either a compulsory or optional part of their degree programme, are responsible for funding travel, accommodation and subsistence costs. These costs vary depending on the location and duration of the placement.

How do I fund my study?

There are different tuition fee and student financial support arrangements for students from Northern Ireland, those from England, Scotland and Wales (Great Britain), and those from the rest of the European Union.

Information on funding options and financial assistance for undergraduate students is available at http://www.qub.ac.uk/Study/Undergraduate/Fees-and-scholarships/.

Scholarships

Each year, we offer a range of scholarships and prizes for new students. Information on scholarships available.

International Scholarships

Information on scholarships for international students, is available at http://www.qub.ac.uk/International/International-students/International-scholarships/.

How and When to Apply

How to Apply

Application for admission to full-time undergraduate and sandwich courses at the University should normally be made through the Universities and Colleges Admissions Service (UCAS). Full information can be obtained from the UCAS website at: www.ucas.com/apply.

When to Apply

UCAS will start processing applications for entry in autumn 2019 from 1 September 2018.

Advisory closing date: 15 January 2019 (18:00).

Late applications are, in practice, accepted by UCAS throughout the remainder of the application cycle, but you should understand that they are considered by institutions at their discretion, and there can be no guarantee that they will be given the same full level of consideration as applications received by the advisory closing date.

Applicants are encouraged to apply as early as is consistent with having made a careful and considered choice of institutions and courses.

The Institution code for Queen's is QBELF and the institution code is Q75.

Further information on applying to study at Queen's is available at: http://www.qub.ac.uk/

Study/ Undergraduate/How-to-apply/

After an offer is made this will be notified to applicants through UCAS. Confirmation will be emailed by the Admissions and Access Service and this communication will also include Terms and Conditions (www.qub.ac.uk/Study/TermsandConditions) which applicants should read carefully in advance of replying to their offer(s) on UCAS Track.

Additional Information for International (non-EU) Students

1. Applying through UCAS

Most students make their applications through UCAS (Universities and Colleges Admissions Service) for full-time undergraduate degree programmes at Queen's. The UCAS application deadline for international students is June 30th 2019.

2. Applying direct

The Direct Entry Application form is to be used by international applicants who wish to apply directly, and only, to Queen's or who have been asked to provide information in advance of submitting a formal UCAS application. Find out more.

3. Applying through agents and partners

The University's in-country representatives can assist you to submit a UCAS application or a direct application. Please consult the Agent List to find an agent in your country who will help you with your application to Queen's University.

Appendix 2: A List of Chemists

A

Richard Abegg, (1869—1910), German chemist, valence theory

Frederick Abel, (1827—1902), English chemist, explosive chemistry

Peter Agre, (1949—), American biochemist and doctor, 2003 Nobel Prize in Chemistry

Georg Agricola, (1494—1555), German scholar known as "the father of mineralogy"

Arthur Aikin, (1773—1855), English chemist and mineralogist

Adrien Albert, (1907—1989), Australian medicinal chemist

Kurt Alder, (1902—1958), German chemist, 1950 Nobel Prize winner in Chemistry, developer of Diels-Alder reaction

Sidney Altman, (1939—), 1989 Nobel Prize winner in Chemistry, RNA catalysis

Christian B. Anfinsen, (1916—1995), 1972 Nobel Prize winner in Chemistry, biochemist

Johan August Arfwedson, (1792—1841), Swedish chemist

Anton Eduard van Arkel, (1893—1976), Dutch chemist

Svante Arrhenius, (1859—1927), Swedish chemist, one of the founders of physical chemistry

Francis William Aston, (1877—1945), 1922 Nobel Prize winner in Chemistry

Amedeo Avogadro, (1776—1856), Italian chemist and physicist

B

Leo Baekeland, (1863—1944), Belgian-American chemist

Adolf von Baeyer, (1835—1917), German chemist, 1905 Nobel Prize winner in Chemistry

Hendrik Willem Bakhuis Roozeboom, (1854—1907), Dutch chemist

Neil Bartlett, (1932—), English/Canadian/American chemist

Sir Derek Barton, (1918—1998), 1969 Nobel Prize winner in Chemistry

Antoine Baum, (1728—1804), French chemist

Karl Bayer, (1847—1904), Austrian chemist

Friedrich Konrad Beilstein, (1838—1906), German-Russian chemist, created Beilstein database

Irina Beletskaya, (1933—), Russian organometallic chemist

Francesco Bellini, (1947—), research scientist doctor in organic chemistry

Paul Berg, (1926—), 1980 Nobel Prize winner in Chemistry

Friedrich Bergius, (1884—1949), 1931 Nobel Prize winner in Chemistry

Marcellin Bertheloi, (1827—1907), French chemist

Claude Louis Berthalle, (1748—1822) French chemist

Jins Jakob, (1779—1848), Swedish chemist

Johannes Martin Bijvoet, (1892—1980), Dutch chemist and crystallographer

Joseph Black, (1728—1799), British chemist

Dale L. Boger, (1953—), American organic and medicinal chemist

Paul Emile Lecoq de Boisbaudran, (1838—1912), French chemist

Jan Boldingh, (1915—2003), Dutch chemist

Alexander Borodin, (1833—1887), Russian chemist & composer

Hans-Joachim Born, German radiochemist

Carl Bosch, (1872—1940), German chemist

Paul D. Boyer, (1918—), American biochemist, 1997 Nobel Prize winner in Chemistry

Robert Boyle, (1627—1691), Irish pioneer of modern chemistry

Henri Braconnot, (1780—1855), French chemist and pharmacist

Sir William Lawrence Bragg, (1890—1971), English X-ray crystallographer and physicist, shared the Nobel Prize in Physics in 1915 with his father Sir William Henry Bragg

Johannes Nicolaus Brensted, (1879—1947), Danish chemist

Herbert C. Brown, (1912—2004), 1979 Nobel Prize winner in Chemistry

Eduard Buchner, (1860—1917), 1907 Nobel Prize winner in Chemistry

Robert Wilhelm Bunsen, (1811—1899), German inventor and chemist

Adolf Butenandt, (1903—1995), 1939 Nobel Prize winner in Chemistry

Aleksandr Butlerov, (1828—1886), Russian chemist

C

Melvin Calvin, (1911—1997), American chemist, 1961 Nobel Prize winner in Chemistry

Georg Ludwig Carius, (1829— 1875), German chemist

Heinrich Caro, (1834—1910), German chemist

Wallace Carothers, (1896—1937), American polymer chemist

Henry Cavendish, (1731—1810), British scientist

Thomas Cech, (1947—), 1989 Nobel Prize winner in Chemistry, RNA research

Martin Chalfie, (1947—), 2008 Nobel Prize winner in Chemistry

Yves Chauvin, (1930—), 2005 Nobel Prize winner in Chemistry, catalysis

Aaron Ciechanover, (1947—), Israeli biochemist, 2004 Nobel Prize winner in Chemistry

Ernst Cohen, (1869—1944), Dutch chemist (murdered in Auschwitz)

Elias James Corey, (1928—), American organic chemist, 1990 Nobel Prize winner in Chemistry, organic chemistry

Robert Corey, (1897—1971), American biochemist

John Cornforth, (1917—), Australian chemist, 1975 Nobel Prize winner in Chemistry

James Crafts, (1839—1917), American chemist, developer of Friedel-Crafts reaction

Donald J. Cram, (1919—2001), American chemist, 1987 Nobel Prize winner in Chemistry

Paul J. Crutzen, (1933—), Dutch chemist, 1995 Nobel Prize winner in Chemistry

Marie Curie, (1867—1934), Polish-born French radiation physicist, 1903 Nobel Prize winner in Physics, 1911 Nobel Prize winner in Chemistry

Pierre Curie, (1859—1906), 1903 Nobel Prize winner in Physics

Robert Curl, (1933—), 1996 Nobel Prize winner in Chemistry

Theodor Curtius, (1857—1928), German chemist

D

John Dalton, (1766—1844), physicist and pioneer of the atomic theory

Curl Peter Henrik Dam, (1895—1976), Danish biochemist, 1943 Nobel Prize winner in Physiology or Medicine

Sumuel J, Danishefsky, (1936—), American organic chemist, natural product total synthesis

Raymond Davis, Jr., (1914—2006), American physical chemist

Humphry Davy, (1778—1829), British inorganic chemist, discoverer of several alkali and alkaline earth metals

Peter Debye, (1884—1966), Dutch physical chemist, 1936 Nobel Prize winner in Chemistry

Johann Deisenhofer, (1943—), 1988 Nobel Prize winner in Chemistry

Sir James Dewar, (1842—1923), British physicist and chemist

Francois Diederich, (1952—), Luxembourg chemist

Otto Diels, (1876—1954), German chemist, 1950 Nobel Prize winner in Chemistry

Edward Doisy, (1893—1986), American biochemist, 1943 Nobel Prize winner in Physiology or Medicine

Davorin Dolar, (1921—2005), Slovenian chemist

David Adriaan van Dorp, (1915—1995), Dutch chemist

Cornelius Drebbel, (1572—1633), Dutch inventor, alchemist and chemist

Jean Baptiste Dumas, (1800—1884), French chemist

Vincent du Vigneaud, (1901—1978), 1955 Nobel Prize winner in Chemistry

E

Paul Ehrlich, (1854—1915), German chemist, 1908 Nobel Prize winner in Physiology or Medicine

Arthur Eichengrün, (1867—1949), Jewish chemist

Manfred Eigen, (1927—), German chemist, 1967 Nobel Prize winner in Chemistry

Mostafa El-Sayed, Egyptian-American physical chemist

Fausto Elhuyar, (1755—1833), Spanish chemist, discoverer of tungsten

Emil Erlenmeyer, (1825—1909), German chemist

Richard R. Ernst, (1933—), 1991 Nobel Prize winner in Chemistry, NMR spectroscopy

Gerhard Ertl, (1936—), German physical winner chemist, 2007 Nobel prize winner in Chemistry

Hans von Euler-Chelpin, (1873—1964), Swedish chemist, 1929 Nobel Prize winner in Chemistry

Henry Eyring, (1901—1981), Mexican-American theoretical chemist

F

Kazimierz Fajans, (1887—1975), Polish-American physical chemist

Michael Faraday, (1791—1867), British physicist and chemist

Hermann von Fehling, (1812—1885), German chemist

John Bennett Fenn, (1917—), 2002 Nobel Prize winner in Chemistry, mass spectrometry

Hermann Emil Fischer, (1852—1919), 1902 Nobel Prize winner in Chemistry, organic chemist, not to be confused with Franz Joseph Emil Fischer (1877—1947)

Ernst Gottfried Fischer, (1754—1831), German chemist

Ernst Otto Fischer, (1918—2007), German chemist, 1973 Nobel Prize winner in Chemistry

Hans Fischer, (1881—1945), German organic chemist, 1930 Nobel Prize winner in Chemistry

Paul Flory, (1910—1985), 1974 Nobel Prize winner in Chemistry, polymer chemist

Rosalind Franklin, (1920—1958), British chemist and crystallographer

Carl Remigius Fresenius, (1818—1897), German chemist

Wilhelm Fresenius, (1913—2004), German chemist, great-grandson of Carl

Charles Friedel, (1832—1899), French developer of Friedel-Crafts reaction

Alexander Naumovich Frumkin, (1895—1976), electrochemist and chemist

Kenichi Fukui, (1918—1998), 1981 Nobel Prize winner in chemistry, quantum chemist

G

Johan Gadolin, (1760—1852), French chemist

Merrill Garnett, (1930—), American biochemist

Joseph Louis Gay-Lussac, (1778—1850), French chemist and physicist

William Giauque, (1895—1982), 1949 Nobel Prize winner in Chemistry

Josiah Willard Gibbs, (1839—1903), American engineer, physical chemist and physicist

Walter Gilbert, (1932—), 1980 Nobel Prize winner in Chemistry

Johann Rudolf Glauber, (1604—1670), Dutch-German alchemist and chemist

Victor Goldschmidt, (1888—1947), Norwegian chemist, "father of modern geochemistry"

Moses Gomberg, (1866—1947), Russian American, "father of free radical chemistry"

David van Goorle, (1591—1612), also called Gorlaeus, Dutch chemist

Thomas Graham, (1805—1869), Scottish chemist

Francois Auguste Victor Grignard, (1871—1935), 1912 Nobel Prize winner in Chemistry, developer of Grignard reagent

Robert H. Grubbs, (1942—), 2005 Nobel Prize winner in Chemistry, catalysis

H

Fritz Haber, (1868—1934), 1918 Nobel Prize winner in Chemistry
Otto Hahn, (1879—1968), 1944 Nobel Prize winner in Chemistry
John Scott Haldane, (1860—1936), British biochemist
Charles Martin Hall, (1863—1914), American chemist
Arthur Harden, (1865—1940), English biochemist, 1929 Nobel Prize winner in Chemistry
Odd Hasset, (1897—1981), Norwegian chemist, 1969 Nobel Prize winner in Chemistry
Charles Hatchett, (1765—1847), English chemist, discovered niobium
Herbert A. Hauptman, (1917—), 1985 Nobel Prize winner in Chemistry
Walter Haworth, (1883—1950), 1937 Nobel Prize winner in Chemistry, organic chemist
Clayton Heathcock, (1936—), American Chemist
Alan J. Heeger, (1936—), 2000 Nobel Prize winner in Chemistry for electrically conductive polymers
Dudley R. Herschbach, (1932—), American physical chemist, 1986 Nobel Prize winner in Chemistry
Avram Hershko, (1937—), 2004 Nobel Prize winner in Chemistry
Charles Herty, (1867—1938), American Chemist
Gerhard Herzberg, (1904—1999), German-Canadian chemist, 1971 Nobel Prize winner in Chemistry, physical Chemist
Germain Henri Hess, (1802—1850), Swiss-born Russian chemist
György von Hevesy, (1885—1966), also called George de Hevesy, Hungarian chemist, 1943 Nobel Prize winner in Chemistry
Jaroslav Heyrovský, (1890—1967), Czech electrochemist, 1959 Nobel Prize winner in Chemistry
Cyril Norman Hinshelwood, (1897—1967), English physical Chemist, 1956 Nobel Prize winner in Chemistry
Dorothy Hodgkin, (1910—1994), 1964 Nobel Prize winner in Chemistry, X-ray crystallography
Jacobus Henricus van't Hoff, (1852—1911), Dutch physical chemist, 1901 Nobel Prize winner in Chemistry
Friedrich Hoffmann, (1660—1742), German physician and chemist
Roald Hoffmann, (1937—), Polish-born American quantum chemist, 1981 Nobel Prize winner in Chemistry
Albert Hofmann, (1906—2008), Swiss chemist, synthesized Lysergic acid diethylamide (LSD)
August Wilhelm von Hofmann, (1818—1892), German organic chemist
Heinrich Hubert Maria Josef Houben, (1875—1940), German organic chemist
Coenraad Johannes van Houten, (1801—1887), Dutch chemist and chocolate maker. invented cocoa powder
Robert Huber, (1937—), German biochemist, 1988 Nobel Prize winner in Chemistry

I

Sir Christopher Kelk Ingold, (1893—1970), English organic chemist

J

Paul Janssen, (1926—2003), Belgian founder of Janssen Pharmaceutica.

Frédéric Joliot-Curie, (1900—1958), French chemist and physicist, 1935 Nobel Prize winner in Chemistry, son-in-law of Marie Curie

Irène Joliot-Curie, (1897—1956), French chemist and physicist, 1935 Nobel Prize winner in Chemistry, daughter of Marie Curie

Louis John Jardino, (1983—), Filipino Chemist

K

Jerome Karle, (1918—), 1985 Nobel Prize winner in Chemistry

Paul Karrer, (1889—1971), 1937 Nobel Prize winner in Chemistry

Karl Wilhelm Gottlob Kastner, (1783—1857), German chemist and natural scientist teacher of chemist Justus von Liebig

August Kekulé, (1829—1896), German organic chemist

John Kendrew, (1917—1997), British biochemist, 1962 Nobel Prize winner in Chemistry

Trevor Kletz, (1922—), British promoter of industrial safety

Aaron Klug, (1926—), British biophysicist, 1982 Nobel Prize winner in Chemistry

Emil Knoevenagel, (1865—1921), organic chemist

William Standish Knowles, (1917—), 2001 Nobel Prize winner in Chemistry, catalysis

Walter Kohn, (1923—), 1998 Nobel Prize winner in Chemistry, American theoretical physicist

Adolph Wilhelm Hermann Kolbe, (1818—1884), organic chemist

Izaak Kolthoff, (1894—1993), Dutch-American chemist, "father of analytical chemistry"

Else Kooi, (1932—2001), Dutch chemist, developed isolation for MOS-transistors

Roger D. Kornberg, (1947—), 2006 Nobel Prize winner in Chemistry for X-ray crystallography

Aleksandra Kornhauser, (1926—), chemist

Harold Kroto, (1939—), English chemist and developer of C_{60}, 1996 Nobel Prize winner in Chemistry

Richard Kuhn, (1900—1967), 1938 Nobel Prize winner in Chemistry

L

Irving Langmuir, (1881—1957), chemist and physicist, 1932 Nobel Prize winner in Chemistry

Paul Lauterbur, (1929—2007), American chemist

Antoine Lavoisier, (1743—1794), French pioneer chemist

Nicolas Leblanc, (1742—1806), French chemist and surgeon

Henri Louis Le Chatelier, (1850—1936), French chemist, famous for devising Le Chatelier's principle to predict the effect of a change in conditions on a chemical equilibrium

Eun Lee, (1946—), Korean organic chemist

Yuan T. Lee, (1936—), 1986 Nobel Prize winner in Chemistry, physical chemist

Jean-Marie Lehn, (1939—), French chemist, 1987 Nobel Prize winner in Chemistry

Luis Federico Leloir, (1906—1987), Argentine biochemist, 1970 Nobel Prize winner in Chemistry

Janez Levec, (1943—), chemist

Gilbert Newton Lewis, (1875—1946), American chemist and first Dean of the Berkeley College of Chemistry

Andreas Libavius, (1555—1616), German doctor and chemist

Willard Libby, (1908—1980), American chemist, 1960 Nobel Prize winner in Chemistry

Justus von Liebig, (1803—1873), German inventor, "father of the fertilizer industry" for his discovery of nitrogen as an essential plant nutrient

Teunis van der Linden, (1884—1965), Dutch chemist, developed insecticide lindane

William Lipscomb, (1919—), 1976 Nobel Prize winner in Chemistry

H. Christopher Longuet-Higgins, British Chemist

Martin Lowry, (1874—1936), British chemist

Ignacy Lukasiewicz, (1802—1882), Polish pharmacist

M

Alan MacDiarmid, (1927—2007), 2000 Nobel Prize winner in Chemistry, polymer chemist

Carolina Henriette Mac Gillavry, (1904—1993), Dutch chemist and crystallographer

Roderick MacKinnon, (1956—), biochemist, 2003 Nobel Prize winner in Chemistry

Pierre Macquer, (1718—1784), French chemist

Rudolph A. Marcus, (born 1923), 1992 Nobel Prize winner in Chemistry

Vladimir Vasilevich Markovnikov, (1838—1904), Russian organic chemist

Tobin J. Marks, American inorganic chemist and material scientist

Alan G. Marshall, American chemist, co-inventor of Fourier transform ion cyclotron resonance (FT-ICR) mass spectrometry

Archer John Porter Martin, (1910—2002), 1952 Nobel Prize winner in Chemistry

Martinus van Marum, (1750—1837), Dutch chemist

Edwin McMillan, (1907—1991), 1951 Nobel Prize winner in Chemistry

Lise Meitner, (1878—1968), German physicist

Dmitri Ivanovich Mendeleev, (1834—1907), Russian chemist, creator of the Periodic Table of Elements

John Mercer, (1791—1866), chemist and industrialist

Robert Bruce Merrifield, (1921—2006), solid-phase chemist, 1984 Nobel Prize winner in Chemistry for polypeptide solid state synthesis

Lothar Meyer, (1830—1895), not to be confused with Viktor Meyer (1848—1897)

August Michaelis, (1847—1916), German biochemist

Hartmut Michel, (1948—), 1988 Nobel Prize winner in Chemistry

Stanley Miller, (1930—), American chemist, best known for the Miller-Urey experiment

Luis E. Miramontes, (1925—2004), co-inventor of the combined oral contraceptive pill

Peter D. Mitchell, (1920—1992), 1978 Nobel Prize winner in Chemistry

William A. Mitchell, (1911—2004), key inventor behind Pop Rocks, Tang, and Kool Whip

Alexander Mitscherlich, (1836—1918), chemist

Henri Moissan, (1852—1907), French chemist, 1906 Nobel Prize winner in Chemistry

Mario J. Molina, (1943—), 1995 Nobel Prize winner in Chemistry

Jacques Monod, (1910—1976), biochemist, 1965 Nobel Prize winner in Physiology or Medicine

Peter Moore, (1939—), American biochemist, Sterling Professor of Chemistry at Yale University

Stanford Moore, (1913—1982), American biochemist, 1972 Nobel Prize winner in Chemistry, ribonuclease research

Henry Gwyn Jeffreys Moseley, (1887—1915), English physicist, discovered Moseley's law

Gerardus Johannes Mulder, (1802—1880), Dutch organic chemist

Robert S. Mulliken, (1896—1986), American physicist, theoretical chemist, 1966 Nobel Prize winner in Chemistry

Kary Mullis, (1944—), 1993 Nobel Prize winner in Chemistry

N

Robert Nalbandyan, (1937—2002), Armenian protein chemist

Giulio Natta, (1903—1979), Italian polymer chemist, 1963 Nobel Prize winner in Chemistry

Costin Nenitescu, (1902—1970), Romanian chemist

Antonio Neri, (1500s—1614), Florentine chemist and glassmaker

Walther Nernst, (1864—1941), German physical chemist, 1920 Nobel Prize winner in Chemistry

John Alexander Reina Newlands, (1837—1898), English analytical chemist

William Nicholson, (1753—1815), English chemist

Kyriacos Costa Nicolaou, American chemist

Alfred Nobel, (1833—1896), Swedish chemist

Ronald George Wreyford Norrish, (1897—1978), British photochemist, 1967 Nobel Prize winner in Chemistry

John Howard Northrop, (1891—1987), 1946 Nobel Prize winner in Chemistry

Ryoji Noyori, (1938—), 2001 Nobel Prize winner in Chemistry, organic chemist

Ralph Nuzzo, (1954—), American surface chemist and materials scientist

O

George Andrew Olah, (1927—), 1994 Nobel Prize winner in Chemistry, organic chemist
Lars Onsager, (1903—1976), physical chemist, 1968 Nobel Prize winner in Chemistry
Luitzen Johannes Oosterhoff, (1907—1974), Dutch chemist
Wilhelm Ostwald, (1853—1932), 1909 Nobel Prize winner in Chemistry

P

Paracelsus, (1493—1541), alchemist
Rudolph Pariser, (1923—), theoretical and organic chemist
Robert G. Parr, (1921—), theoretical chemist
Louis Pasteur, (1822—1895), French biochemist
Linus Pauling, (1901—1994), 1954 Nobel Prize winner in Chemistry, 1962 Nobel Peace Prize winner
Charles J. Pedersen, (1904—1989), 1987 Nobel Prize winner in Chemistry
Eugène-Melchior Péligot, (1811—1890), French chemist who isolated the uranium metal
William Henry Perkin, (1838—1907), British organic chemist and inventor of mauveine (dye)
William Henry Perkin, Jr., (1860—1929), British organic chemist, son of Sir William Henry Perkin
Max Perutz, (1914—2002), 1962 Nobel Prize winner in Chemistry, biochemist
David Andrew Phoenix, (1966—), Biochemist
Roy J. Plunkett, (1910—1994), discoverer of Teflon
John Charles Polanyi, (1929—), Canadian physical chemist, 1986 Nobel Prize winner in Chemistry
John A. Pople, (1925—2004), theoretical chemist, 1998 Nobel Prize winner in Chemistry
George Porter, (1920—2002), British photochemist, 1967 Nobel Prize winner in Chemistry
Fritz Pregl, (1869—1930), chemist, 1923 Nobel Prize winner in Chemistry
Vladimir Prelog, (1906—1998), 1975 Nobel Prize winner in Chemistry
Joseph Priestley, (1733—1804), English theologian, natural philosopher, and educator who published over 150 works. Discovered oxygen and isolated it in its gaseous state, although Carl Scheele and Antoine Lavoisier also have claim to the discovery
Ilya Prigogine, (1917—2003), 1977 Nobel Prize winner in Chemistry, physical chemist

Q

Ğilem Qamay, (1901—1970), Soviet chemist

R

William Ramsay, (1852—1916), Scottish chemist, 1904 Nobel Prize winner in Chemistry

Henry Rapoport, American chemist, UC Berkeley

William Sage Rapson, South African Chemist and co-author of *Gold Usage*

Julies Rebek, Hungarian-American chemist

Jan Reedijk, (1943—), Dutch inorganic chemist

Henri Victor Regnault, (1810—1878), French chemist and physicist

Tadeus Reichstein, (1897—1996), chemist, 1950 Nobel Prize winner in Physiology or Medicine

Rhazes (Razi), (865—925), Iranian chemist

Stuart A. Rice, (1932—), physical chemist

Theodore William Richards, (1868—1928), 1914 Nobel Prize winner in Chemistry

Paul Wilhelm Richter, (1946—), South Africa, bioceramics

Ellen Swallow Richards, (1842—1911), industrial and environmental chemist

Jeremias Benjamin Richards, (1762—1807), German chemist

Nikolaus Riehl, (1901—1990), industrial nuclear chemist

Andrés Manuel del Río, (1764—1849), Spanish-Mexico geochemist, discovered vanadium

Robert Robinson, (1886—1975), 1947 Nobel Prize winner in Chemistry

Hillar Rootare, (1928—), Estonian-American physical chemist

Irwin Rose, (1926—), American biochemist, 2004 Nobel Prize winner in Chemistry

Guillaume-Francois Rouelle, (1703—1770), French chemist

H. M. Rouell, (1718—1779), French chemist

Frank Sherwood Rowland, (1927—), 1995 Nobel Prize winner in Chemistry

Daniel Rutherford, (1749—1819), Scottish chemist

Ernest Rutherford, (1871—1937), New Zealand born chemist and nuclear physicist. Discovered the Proton. 1908 Nobel Prize winner in Chemistry

Leopold Ruzicka (Lavoslav Ružička), (1887—1976), 1939 Nobel Prize winner in Chemistry

S

Paul Sabatier, (1854—1941), 1912 Nobel Prize winner in Chemistry

Maks Samec, (1844—1889), Slovenian chemist

Frederick Sanger, (1918—), British biochemist, 1958 and 1980 Nobel Prize winner in Chemistry

Carl Wilhelm Scheele, (1742—1786), Swedish chemist, discovered numerous elements

Stuart L. Schreiber, (1956—), American chemist, a pioneer in a field of chemical biology

Richard R. Schrock, (1945—), 2005 Nobel Prize winner in Chemistry for catalysis

Peter Schultz, (1956—), American chemist

Glenn T. Seaborg, (1912—1999), 1951 Nobel Prize winner in Chemistry

Nils Gabriel Sefström, (1787—1845), chemist

Francesco Selmi, (1817—1881), Italian chemist

Nikolay Nikolayevich Semyonov, (1896—1986), physicist and chemist, 1956 Nobel Prize winner in Chemistry

K. Barry Sharpless, (1941—), 2001 Nobel Prize in Chemistry

Patsy O. Sherman, (1930—), 12 US Patents

Osamu Shimomura, (1928—), 2008 Nobel Prize winner in Chemistry

Hideki Shirakawa, (1936—), 2000 Nobel Prize winner in Chemistry, conductive polymers

Alexander Shulgin, (1925—), a pioneer researcher in Psychopharmacology and Entheogens

Salimuzzaman Siddiqui, (1897—1994), Pakistani chemist, pioneer in natural products chemistry

Oktay Sinanoglu, (1935—), Turkish chemist

Jens Christian Skou, (1918—), 1997 Nobel Prize winner in Chemistry

Richard Smalley, (1943—2005), 1996 Nobel Prize winner in Chemistry, developer of C_{60}

Michael Smith, (1932—2000), 1993 Nobel Prize winner in Chemistry

Ascanio Sobrero, (1812—1888), Italian chemist, discoverer of nitroglycerin

Frederick Soddy, (1877—1956), British chemist, 1921 Nobel Prize winner in Chemistry

Susan Solomon, American atmospheric chemist

Ernest Solvay, (1838—1922), Belgian chemist and industrialist

S.P.L. Sorensen, (1868—1939), Danish chemist

Wendell Meredith Stanley, (1904—1971), 1946 Nobel Prize winner in Chemistry

Branko Sianovnik, (1938—), chemist

Hermann Staudinger, (1881—1965), polymer chemist, 1953 Nobel Prize winner in Chemistry

William Howard Stein, (1911—1980), 1972 Nobel Prize winner in Chemistry

Alfred Stock, (1876—1946), German inorganic chemist, a pioneer in research on the hydrides of boron and silicon, coordination chemistry, mercury, and mercury poisoning.

Fraser Stoddart, (1945—), Scottish chemist, a pioneer in the field of the mechanical bond

James B. Sumner, (1887—1955), 1946 Nobel Prize winner in Chemistry

Theodor Svedberg, (1884—1971), 1926 Nobel Prize winner in Chemistry

Joseph Swan, (1828—1914), English physicist, chemist & inventor

Richard Laurence Millington Synge, (1914—1994), 1952 Nobel Prize winner in Chemistry

T

Koichi Tanaka, (1959—), 2002 Nobel Prize winner in Chemistry, mass spectrometry

Henry Taube, (1915—2005), inorganic chemist, 1983 Nobel Prize winner in Chemistry

Richard Taylor, (1965—), Professor of Organic Chemistry, University of York

Arne Tiselius, (1902—1971), Swedish biochemist, 1948 Nobel Prize winner in Chemistry, electrophoresis

Max Tishler, (1906—1989), 1970 Priestley Medal

Miha Tisler, (1926—), chemist

Alexander R. Todd, (1907—1997), 1957 Nobel Prize winner in Chemistry

U

Harold Clayton Urey, (1893—1981), American chemist, 1934 Nobel Prize winner in Chemistry for isotopes research

V

Lauri Vaska, (1925—), Estonian-American chemist
Evert Johannes Willem Verweij, (1905—1981), Dutch chemist
Artturi Ilmari Virtanen, (1895—1973), chemist, Nobel Prize winner, laureate
Max Volmer, (1885—1965), German chemist
Alessandro Volta, (1745—1827), electrochemist, invented the Voltaic Cell

W

Johannes Diderik van der Waals, (1837—1923), physical chemist
Sir James Walker, (1863—1935), Scottish physical chemist
John E. Walker, (1941—), 1997 Nobel Prize winner in Chemistry, biochemist
Otto Wallach, (1847—1931), 1910 Nobel Prize winner in Chemistry
Alfred Werner, (1866—1919), 1913 Nobel Prize winner in Chemistry
Peter Jaffrey Wheatley, (1921—1997), British physical chemist
George M. Whitesides, (1939—), American chemist
Heinrich Otto Wieland, (1877—1957), German chemist, 1927 Nobel Prize winner in Chemistry
Harvey W, Wiley, (1844—1930), American chemist, pure food & drug advocate
Sir Geoffrey Wilkinson, (1921—1996), 1973 Nobel Prize winner in Chemistry, organic chemist
Richard Willstätter, (1872—1942), 1915 Nobel Prize winner in Chemistry
Adolf Otto Reinhold Windaus, (1876—1959), 1928 Nobel Prize winner in Chemistry
Günter Wirths, (1911—), German nuclear chemist
Georg Wittig, (1897—1987), German organic chemist, 1979 Nobel Prize winner in Chemistry
Friedrich Wöhler, (1800—1882), German chemist
William Hyde Wollaston, (1766—1828), English chemist
Robert B. Woodward, (1917—1979), organic chemist, 1965 Nobel Prize winner in Chemistry
Charles-Adolphe Wurtz, (1817—1884), organic chemist, developer of Wurtz reaction
Kurt Wüthrich, (1938—), 2002 Nobel Prize winner in Chemistry, biochemist

Y

Sabir Yunusov, (1909—1995), Soviet chemist in alkaloids

Z

Ahmed H. Zewail, (1946—), Egyptian, 1999 Nobel Prize winner in Chemistry for his work on femtochemistry

Karl Ziegler, (1898—1973), German polymer chemist, 1963 Nobel Prize winner in Chemistry

Richard Adolf Zsigmondy, (1865—1929), colloidal chemist, 1925 Nobel Prize winner in Chemistry

Appendix 3: A List of Chemical Engineers

A

Mohammed Al Mady, executive president of SABIC. SABIC

Mukesh Ambani, Chairman and Managing Director of Reliance Industries. Reliance Industries

Ramani Ayer, CEO of the Hartford

B

Jay Bailey, (1944—2001), pioneering work in metabolic engineering, ETH Zurich

Samuel Bodman, former Chairman, CEO of Cabot Corporation, United States Secretary of Energy (2005—2009)

Carl Bosch, from 1908 to 1913 developed the Haber-Bosch process together with Fritz Haber. His other notable work was for the introduction of high pressure chemistry. 1931 Nobel Prize winner in Chemistry

Henry Bessemer, (1813—1898), invented Bessemer process for manufacturing steel independent inventor

C

John G. Collier, two-phase flow and nuclear power expert, head of UKAEA and CEGB. UKAEA

John Coulson, co-author of what became the standard UK textbook set; Coulson & Richardson's Chemical Engineering, Newcastle University, UK

D

Donald A. Dahlstrom, (1920—2005), inventor of the Hydrocyclone and related correlations. Elected to National Academy of Engineering in 1975 for "Contributions to liquid-solids separation processes in mineral recovery and waste disposal". University of Utah

Zara Salim Davidson, wife of the Raja Muda (Crown Prince) of Perak

George E. Davis, regarded as the "founding father" of chemical engineering

Nguyet Anh Duong, a Vietnamese-American, she assisted in creating a new weapon called the Thermobaric weapon in support of Operation Enduring Freedom to assault tunnels and caves being used as hideouts by the Taliban in Afghanistan.

John Drosdick, (1943-), Chairman and CEO of Sunoco, Inc.

E

Lawrence B. Evans, CEO of Rive Technology, Inc. Contributed significantly to commercial development of ASPEN process simulation technology. Elected to National Academy of Engineering in 2001 for "leadership in the development and application of integrated systems for modeling, simulation, and optimization of industrial chemical processes". The AIChE Lawrence B. Evans Award in Chemical Engineering Practice is named for him. Rive Technology, Inc.

F

James R. Fair, noted for research into separation processes, especially distillation. Elected to National Academy of Engineering in 1974 for "contributions to mass transfer technology and computer simulation of chemical processes". Received award from AIChE in 1985 for outstanding chemical engineering contributions. University of Texas

Richard Felder, multiple award-winning engineering educator. North Carolina State University

Ian Fells, energy expert and popular science broadcaster. Newcastle University, UK

Merrill R. Fenske, (1904—1971), professor of chemical engineering. Developed the Fenske equation, to calculate the minimum reflux needed to separate two binary components by continuous distillation. Pennsylvania State University

John Bennett Fenn, 2002 Nobel Prize winner in Chemistry. Virginia Commonwealth University

G

Rafiqul Gani, professor of chemical engineering at DTU, editor-in-chief of *Computers & Chemical Engineering* journal. Technical University of Denmark (DTU)

Clifton C Garvin, Chairman and CEO, Exxon Eugenio Garza Lagüera Chairman and CEO. FEMSA

Edwin R. Gilliland, (1909—1963), professor of chemical engineering. Noted for research into distillation and fluid bed catalytic cracking. Namesake of the Gilliland correlation used in designing distillation columns. Elected to the National Academy of Sciences in 1948, and the National Academy of Engineering in 1965. MIT

Roberto Goizueta, former Chairman and Chief Executive of Coca-Cola. Coca-Cola

Robert W. Gore, the inventor of Gore-Tex

William Sealey Gosset, Brewer and statistician

Andrew Grove, Intel Chief Executive Officer. Intel

Pierre Gy, developed theory of sampling

H

Fritz Haber, (1868—1934), received Nobel Prize in Chemistry 1918 for the fixation of nitrogen from the air, the Haber process; also developed chemical warfare agents for the German government during World War I. Received the Rumford Medal in 1932 for "… the outstanding importance of his work in physical chemistry, especially in the application of thermodynamics to chemical reactions". University of Karlsruhe (TH)

Vladimir Haensel, (1914—2002), invented the "Platforming" (Platinum Reforming) process, which led to the production of low cost high octane gasoline, and contributed to development of catalytic converters for automobiles. University of Massachusetts Amherst

Douglas Patrick Harrison, carried out research for DOE Vision 21 project as well research to remove CO_2 from stack gas of coal-fired power generators and for production of pure Hydrogen from gasification of coal. Louisiana State University

Fred Hassan, CEO and Chairman of the Board of Schering-Plough Corporation; former Chairman and CEO for the Pharmacia Corporation. Schering-Plough

Csaba Horvath, (1930—2004), considered as one of the pioneers of modem separation science. Yale University

Hu Tsu Tau Richard, former Minister for Finance (Singapore) (1985—2001)

I

Dan Itse, US politician and holder of four patent on low-emission technology. Worcester Polytechnic-Institute

J

Rakech Jain, integrated bioengineeing with tumor biology and imaging gene expression and functions *in vivo* for drug delivery in tumors. Harvard Medical School

Mac Jemison, science mission specialist on the Space Shuttles Endeavour and first black woman in space. NASA

K

Donald Q. Kern, Chemical engineering professor at Case Institute of Technology (now Case Western Reserve University). Author of *Process Heat Transfer*, a major reference for designers of heat transfer equipment. Namesake of the AIChE Donald Q. Kern award. Case Western Reserve University

Henry Z. Kister, author of books and articles on distillation. Cited by *Chemical Engineering* magazine for personal achievement in chemical engineering. C. F. Braun (now KBR (Company). Fluor Corp

Trevor Kletz, author of books dealing with chemical engineering safety at Loughborough University

Riki Kobayashi, Professor emeritus of chemical engineering. Notable for research in phase equilibria and physical properties related to natural gas processing. Elected a member of the National Academy of Engineering in 1995 "…for advances in the knowledge and measurement of the thermodynamic and transport properties of natural gas liquids and gas hydrates". Rice University

Charles G. Koch, CEO of Koch Industries. Koch Industries

David H. Koch, executive vice-president of Koch Industries. Koch Industries

L

Robert Langer, Tissue Engineering and Controlled- Release Drug Delivery pioneer. MIT

Frank Lees, author and pioneer of Loss Prevention in the process industries at Loughborough University

Warren K. Lewis, (1882—1975), American chemical engineering professor; played a role in defining the field of chemical engineering during its early development. Co-developer (with E. R. Gilliland) of Houdry process for petroleum refining. MIT

Arthur Dehon Little, (1863—1935), consultant and co-founder, with William Walker, of "Arthur D. Little, Inc", a major consulting firm. Arthur D. Little, Inc.

M

John F. MacGregor, use of latent variables in industrial processes. McMaster University

John J. McKetta, Professor emeritus of chemical engineering at University of Texas. Editor of *Encyclopedia of Chemical Processing and Design*. Emeritus member of the National Academy of Engineering. Elected 1970 for "…enlightening experiments on fluid behavior and the director of technical publications". University of Texas

Victor Mills, (1897—1997), inventor of the disposal diaper. Procter & Gamble

Luis E. Miramontes, co-inventor of the progestin used in one of the first oral contraceptives UNAM. Mexico

Mario Molina, 1995 Nobel Prize winner in Chemistry. UCSD

Frank Morton, namesake of Frank Morton Sports. Day Birmingham University/Manchester University

N

Dudley Maurice Newitt, (1894—1980), scientific director of Special Operations Executive developing gadgets for spies during World War II —a real life Q (James Bond). Received Rumford Medal in 1962 for "…his distinguished contributions to chemical engineering". Imperial College

O

Lars Onsager, 1968 Nobel Prize winner in Chemistry. Yale University

Adam Osborne, (1939—2003), introduced the first ever portable computer in 1981, the same year IBM launched the personal computer. Osborne Computer Corporation

Donald Othmer, (1904—1995), co-creator of the Kirk-Othmer Encyclopedia of Chemical Technology Polytechnic. University of New York

David J. O'Reilly, Chairman and CEO of Chevron. Corporation Chevron Corporation

P

Ding-Yu Peng, co-discoverer of the Peng-Robinson equation of state

Martin Perl, 1995 Nobel Prize winner in Physics. University of Michigan

Robert H. Perry, author of *Handbook* in 1934, now *Perry's Chemical Engineers' Handbook*. University of Oklahoma

Nicholas A. Peppas, pioneer in drug delivery, biomaterials, hydrogels and nanobiotechnology. University of Texas at Austin

Q

K. B. Quinan, explosive manufacturing expert in World War I and first vice-president of the Institution of Chemical Engineers. De Beers

R

Lee Raymond, ExxonMobil Chairman and Chief Executive Officer. ExxonMobil

George Maxwell Richards, President of Trinidad and Tobago

Jack Richardson, co-wrote the textbook which became UK standard Coulson & Richardson's Chemical Engineering. Swansea University

Norbert Rillieux, inventor who is most noted for developing the process that turned sugar from a luxury to a common commodity

Margaret Hutchinson Rousseau, designed the first commercial penicillin production plant

S

Albert Sacco, US astronaut. Worcester Polytechnic Institute

Lanny Schmidt, surface science and detailed chemistry. University of Minnesota

Waldo L. Semon, inventor who patented more than 116 inventions, including polyvinyl chloride (PVC). B.F. Goodrich

Thomas Kilgore Sherwood, (1903—1976), American chemical engineering professor, after whom the Sherwood number in mass transfer is named. Published the textbook, *Absorption and Extraction* in 1937. Was a founding member of the National Academy of Engineering. MIT

Jack Steinberger 1988 Nobel Prize winner in Physics. Columbia University

T

Frank M. Tiller, (1917—2006), M. D. Anderson Distinguished Professor at University of Houston. Noted for research in solid-liquid separation and a founder of the American Filtration and Separation Society. University of Houston

U

A. J. V. Underwood, developed a correlation to estimate the minimum reflux required to produce a specific concentration of a feed component in the distillate from a specified concentration in the feed. The Underwood equation is also known as the Underwood-Fenske equation. University College of London

Lewis Urry, invention of long-lasting alkaline batteries. Eveready Battery Co.

W

Jack Welch, former Chairman and Chief Executive Officer. General Electric

Nathaniel C. Wyeth, inventor of PET plastic bottles. DuPont

Z

F. A. Zenz, received award from AIChE in 1985 for outstanding chemical engineering contributions. Has published extensively on the topic of fluidized particles, including fluidized beds and pneumatic conveying

Appendix 4: A List of Vocabulary in Chemistry and Chemical Engineering

atom	原子
element	元素
ion	离子
anion	阴离子
cation	阳离子
electron	电子
neutral	中性的
proton	质子
atomic nucleus	原子核
nucleon	核子
nuclide symbol	核素符号
nuclide	核素
isotope	同位素

Appendix 5: Element Names and Their Pronunciation

Atomic Number	Element Symbol	English Name	Pronunciation	中文名称	汉字读音
1	H	hydrogen	['haɪdrədʒən]	氢	qīng
2	He	helium	['hɪːljəm, -lɪam]	氦	hài
3	Li	lithium	['lɪθɪəm]	锂	lǐ
4	Be	beryllium	[bə'rɪljəm]	铍	pí
5	B	boron	['bɔːrən]	硼	péng
6	C	carbon	['kaːbən]	碳	tàn
7	N	nitrogen	['naɪtrədʒən]	氮	dàn
8	O	oxygen	['ɔksɪdʒən]	氧	yǎng
9	F	fluorine	['flɔːrɪːn]	氟	fú
10	Ne	neon	['nɪːɒn]	氖	nǎi
11	Na	sodium	['səudjəm,-dɪəm]	钠	nà
12	Mg	magnesium	[mæg'nɪːzjəm]	镁	měi
13	Al	aluminium	[ælju'mɪnɪəm]	铝	lǚ
14	Si	silicon	['sɪlɪkən]	硅	guī
15	P	phosphorus	['fɔsfərəs]	磷	lín
16	S	sulphur	['sʌlfə]	硫	liú
17	Cl	chlorine	['klɔːrɪːn]	氯	lǜ
18	Ar	argon	['aːgɔn]	氩	yà
19	K	potassium	[pə'tæsjəm]	钾	jiǎ
20	Ca	calcium	['kælsɪəm]	钙	gài
21	Sc	scandium	[s'kændɪəm]	钪	kàng
22	Ti	titanium	[taɪ'teɪnjəm]	钛	tài
23	V	vanadium	[və'neɪdɪəm,-djəm]	钒	fán
24	Cr	chromium	['krəumjəm]	铬	gè
25	Mn	manganese	[mæŋgə'nɪːz]	锰	měng
26	Fe	iron	['aɪən]	铁	tiě
27	Co	cobalt	[kə'bɔːlt,'kəubɔːlt]	钴	gǔ
28	Ni	nickel	['nɪkl]	镍	niè
29	Cu	copper	['kɔpə]	铜	tóng
30	Zn	zinc	[zɪŋk]	锌	xīn

续表

Atomic Number	Element Symbol	English Name	Pronunciation	中文名称	汉字读音
31	Ga	gallium	['gælɪəm]	镓	jiā
32	Ge	germanium	[dʒɑːˈmeɪnɪəm]	锗	zhě
33	As	arsenic	[ˈɑːsənɪk]	砷	shēn
34	Se	selenium	[sɪˈliːnɪəm,-njəm]	硒	xī
35	Br	bromine	[ˈbrəumiːn]	溴	xiù
36	Kr	krypton	[ˈkrɪptɒn]	氪	kè
37	Rb	rubidium	[ruːˈbɪdɪəm]	铷	rú
38	Sr	strontium	[ˈstrɒnʃɪəm]	锶	sī
39	Y	yttrium	[ˈɡɪtrɪəm]	钇	yǐ
40	Zr	zirconium	[zɜːˈkəʊnɪəm]	锆	gào
41	Nb	niobium	[naɪˈəʊbɪəm]	铌	ní
42	Mo	molybdenum	[məˈlɪbdənəm]	钼	mù
43	Tc	technetium	[tekˈniːʃɪəm]	锝	dé
44	Ru	ruthenium	[ruːˈθiːnɪəm]	钌	liǎo
45	Rh	rhodium	[ˈrəʊdɪəm]	铑	lǎo
46	Pd	palladium	[pəˈleɪdɪəm]	钯	bǎ
47	Ag	silver	[ˈsɪlvə(r)]	银	yín
48	Cd	cadmium	[ˈkædmɪəm]	镉	gé
49	In	indium	[ˈɪndɪəm]	铟	yīn
50	Sn	tin	[tɪn]	锡	xī
51	Sb	antimony	[ˈæntɪmənɪ]	锑	tī
52	Te	tellurium	[teˈljʊərɪəm]	碲	dì
53	I	iodine	[ˈaɪədiːn]	碘	diǎn
54	Xe	xenon	[ˈzenɒn]	氙	xiān
55	Cs	caesium	[ˈsiːzɪəm]	铯	sè
56	Ba	barium	[ˈbeərɪəm]	钡	bèi
57	La	Lanthanum	[ˈlænθənəm]	镧	lán
58	Ce	cerium	[ˈsɪərɪəm]	铈	shì
59	Pr	praseodymium	[ˌpreɪzɪəʊˈdɪmɪəm]	镨	pǔ
60	Nd	neodymium	[ˌniːəʊˈdɪmɪəm]	钕	nǚ
61	Pm	promethium	[prəˈmiːθɪəm]	钷	pǒ
62	Sm	samarium	[səˈmeərɪəm]	钐	shān
63	Eu	europium	[jʊərˈəʊpɪəm]	铕	yǒu
64	Gd	gadolinium	[ˌɡædəˈlɪnɪəm]	钆	gá

续表

Atomic Number	Element Symbol	English Name	Pronunciation	中文名称	汉字读音
65	Tb	terbium	['tɜ:bɪəm]	铽	tè
66	Dy	dysprosium	[dɪs'prəʊzɪəm]	镝	dī
67	Ho	holmium	['həʊlmɪəm]	钬	huǒ
68	Er	erbium	['ɜ:bɪəm]	铒	ěr
69	Tm	thulium	['θu:lɪəm]	铥	diū
70	Yb	ytterbium	[ɪ'tɜ:bɪəm]	镱	yì
71	Lu	lutetium	[lu:'tɪ:ʃɪəm]	镥	lǔ
72	Hf	hafnium	['hæfnɪəm]	铪	hā
73	Ta	tantalum	['tæntələm]	钽	tǎn
74	W	tungsten	['tʌŋstən]	钨	wū
75	Re	rhenium	['rɪ:nɪəm]	铼	lái
76	Os	osmium	['ɒzmɪəm]	锇	é
77	Ir	iridium	[ɪ'rɪdɪəm]	铱	yī
78	Pt	platinum	['plætɪnəm]	铂	bó
79	Au	gold	[gəʊld]	金	jīn
80	Hg	mercury	['mɜ:kjərɪ]	汞	gǒng
81	Tl	thallium	['θælɪəm]	铊	tā
82	Pb	lead	[lɪ:d]	铅	qiān
83	Bi	bismuth	['bɪzməθ]	铋	bì
84	Po	polonium	[pə'ləʊnɪəm]	钋	pō
85	At	astatine	['æstətɪ:n]	砹	ài
86	Rn	radon	['reɪdɒn]	氡	dōng
87	Fr	francium	['frænsɪəm]	钫	fāng
88	Ra	radium	['reɪdjəm]	镭	léi
89	Ac	actinium	[æk'tɪnɪəm]	锕	ā
90	Th	thorium	['θɔ:rɪəm]	钍	tǔ
91	Pa	protactinium	[prəʊtəʊæk'tɪnɪəm]	镤	pú
92	U	uranium	[jʊə'reɪnɪəm]	铀	yóu
93	Np	neptunium	[nep'tju:nɪəm]	镎	ná
94	Pu	plutonium	[plu:'təʊnɪəm]	钚	bù
95	Am	americium	[æmə'rɪʃɪəm]	镅	méi
96	Cm	curium	['kjʊərɪəm]	锔	jú
97	Bk	berkelium	[bɜ:'kɪ:lɪəm]	锫	péi
98	Cf	californium	[ˌkælɪ'fɔ:nɪəm]	锎	kāi

续表

Atomic Number	Element Symbol	English Name	Pronunciation	中文名称	汉字读音
99	Es	einsteinium	[aɪnˈstaɪnɪəm]	锿	āi
100	Fm	fermium	[ˈfɜːmɪəm]	镄	fèi
101	Md	mendelevium	[ˌmendəˈliːvɪəm]	钔	mén
102	No	nobelium	[nəʊˈbiːlɪəm]	锘	nuò
103	Lr	lawrencium	[lɒˈrensɪəm]	铹	láo
104	Rf	rutherfordium	[ˌrʌðəˈfɔːdɪəm]	𬬻	lú
105	Db	dubnium	[ˈdʌbnɪəm]	𬭊	dù
106	Sg	seaborgium	[sɪˈbɔːgɪəm]	𬭳	xǐ
107	Bh	bohrium	[ˈbɔːrɪəm]	𬭛	bō
108	Hs	hassium	[ˈhæsɪəm]	𬭶	hēi
109	Mt	meitnerium	[maɪtˈnɪərɪəm]	鿏	mài
110	Ds	darmstadtium	[ˈdɑːmʃtætɪəm]	𫟼	dá
111	Rg	roentgenium	[ˌrɒntˈgɪːnɪəm]	𬬭	lún

Appendix 6: Bilingual List of common Analytical Instrument and Methods

English Name	中文名称
Aging Property Tester	老化性能测定仪
Amino Acid Analyzer	氨基酸组成分析仪
Analyzer for Clinic Medicine Concentration	临床药物浓度仪
Atomic Absorption Spectroscopy (AAS)	原子吸收光谱仪
Atomic Emission Spectrometer (AES)	原子发射光谱仪
Atomic Fluorescence Spectroscopy (AFS)	原子荧光光谱仪
Automatic Analyzer for Microbes	微生物自动分析系统
Automatic Titrator	自动滴定仪
Bechtop	超净工作台
Biochemical Analysis	生物技术分析
Biochemical Analyzer	生化分析仪
Bio-reactor	生物反应器
Blood-gas Analyzer	血气分析仪
Centrifuge	离心机
Chemiluminescence Apparatus	化学发光仪
CHN Analysis	环境成分分析仪
CO_2 Incubators	二氧化碳培养箱
Combustion Property Tester	燃烧性能测定仪
Conductivity Meter	电导仪
Constant Temperature Circulator	恒温循环泵
Differential Scanning Calorimetry (DSC)	差示扫描量热法
Differential Thermal Analysis (DTA)	差示热分析法
Direct Current Plasma Emission Spectrometer (DCP)	直流等离子体发射光谱仪
DNA Sequencers	DNA 测序仪
DNA Synthesizer	DNA 合成仪
Electrical Property Tester	电性能测定仪
Electro Microscopy	电子显微镜
Electrolytic Analyzer	电解质分析仪
Electron Energy Disperse Spectroscopy	电子能谱仪
Electrophoresis	电泳

续表

English Name	中文名称
Mass Spectrometer (MS)	质谱
Liquid Chromatograph-Mass spectrometer (LC-MS)	液相色谱-质谱联用仪
ICP-MS	ICP-质谱联用仪
Electrophoresis System	电泳仪
ELISA	酶标仪
Energy Disperse Spectroscopy (EDS)	能谱仪
Fermenter	发酵罐
Flow Cytometer	流式细胞仪
Fraction Collector	部分收集器
Freeze Drying Equipment	冻干机
FT-IR Spectrometer (FTIR)	傅立叶变换红外光谱仪
FT-Raman Spectrometer (FTIR-Raman)	傅立叶变换拉曼光谱仪
Gas Chromatograph (GC)	气相色谱仪
Gas Chromatograph-Mass spectrometer (GC-MS)	气相色谱-质谱联用仪
Gel Permeation Chromatograph (GPC)	凝胶渗透色谱
Hematocyte Counter	血球计数器
High Performance Liquid Chromatography (HPLC)	高效液相色谱仪
Hybridization Oven	分子杂交仪
Inductive Coupled Plasma Emission Spectrometer (ICP)	电感耦合等离子体发射光谱仪
Infrared Spectrometer (IR)	红外光谱仪
Instrument for Nondestructive Testing	无损检测仪
Inverted Microscope	倒置显微镜
Ion Chromatograph	离子色谱
Isotope X-Ray Fluorescence Spectrometer	同位素X荧光光谱仪
Mechanical Property Tester	机械性能测定仪
Metal/material Elemental Analysis	金属/材料元素分析仪
Metallurgical Microscopy	金相显微镜
Microwave Inductive Plasma Emission Spectrometer (MIP)	微波等离子体光谱仪
Nuclear Magnetic Resonance Spectrometer (NMR)	核磁共振波谱仪
Optical Microscopy	光学显微镜
Optical Property Tester	光学性能测定仪
Particle Size Analyzer	粒度分析仪
PCR Amplifier; Instrument for Polymerase Chain Reaction	DNA扩增仪
Peptide Synthesizer	多肽合成仪
Physical Property Analysis	物性分析

续表

English Name	中文名称
Polarograph	极谱仪
Protein Sequencer	氨基酸测序仪
Rheometer	流变仪
Scanning Electron Microscope (SEM)	扫描电镜
Scanning Probe Microscopy	扫描探针显微镜
Sensors	传感器
Shaker	摇床
Size Exclusion Chromatograph (SEC)	体积排阻色谱
Surface Analyzer	表面分析仪
Thermal Analyzer	热分析仪
Thermal Physical Property Tester	热物性能测定仪
Thermogravimetric Analysis (TGA)	热重分析法
Transmitance Electron Microscope (TEM)	透射电镜
Ultrahigh Purity Filter	超滤器
Ultra-low Temperature Freezer	超低温冰箱
Ultrasonic Cell Disruptor	超声破碎仪
Ultraviolet Detector	紫外检测仪
Ultraviolet Lamp	紫外观察灯
Urine Analyzer	尿液分析仪
Viscometer	黏度计
Voltammerter	伏安仪
Water Test Kits	水质分析仪
X-Ray Diffractomer (XRD)	X射线衍射仪
X-Ray Fluorescence Spectrometer (XRF)	X射线荧光光谱仪

Appendix 7: Terms Used in General and Inorganic Chemistry

Abrasive

A very hard heat-resistant material (such as diamond, corundum, boron carbide, ...) that is used to grind the edges or rough surfaces of an object.

Abrasive materials

Hard materials used to shape other materials by grinding or abrading action.

Absolute temperature scale

A temperature scale in which the lowest temperature that can be attained theoretically is zero.

Absolute zero (0 K)

Title temperature at which the volume of an ideal gas becomes zero; a theoretical coldest (lowest) temperature that can be approached but never reached.

Acid

The most common definition, due to Arrhenius, is that an acid is a chemical which produces hydrogen ions when dissolved in water.

Acid dissociation constant; acid ionization constant

The equilibrium constant for the ionization of a weak acid in water, giving the hydronium ion and the conjugate base of the acid.

Acid rain; acid deposition

Rainwater with an acidic pH as a result of air pollution by sulfur dioxide and nitrogen oxides.

Acid-base indicator

A dye used to distinguish between acidic and basic solutions by means the color changes it undergoes in these solutions.

Acid-base reaction

A chemical reaction between an acid and a base.

Acidify

To make or become acidic; convert into an acid. In chemistry, the term is used when acid is added gradually to a solution to bring the pH down below 7.

Acids and bases (definitions)

Two interrelated classes of chemical compounds, the precise definition of which have varied considerably with the development of chemistry.

Activated charcoal; activated carbon

An amorphous form of carbon.

Activation energy

The minimum energy of collision required for two molecules to react.

Active site; active center

Those sites for adsorption which are the effective sites for a particular heterogeneous catalytic reaction.

Actual yield

Amount of a specified pure product actually obtained from a given reaction.

Adenosine diphosphate (ADP)

A coenzyme and an important intermediate in cellular metabolism as the partially dephosphorylated form of adinosinetriphosphate.

Adenosinetriphosphate (ATP)

A coenzyme and one of the most important compound in the metabolism of all organisms.

Adhesive

A material capable of fastening two other materials together by means of surface attachment.

Aerobic

A processor event that requires molecular oxygen.

Affinity

The tendency of a molecule to associate with another ordered structure.

Air pollution

The contamination of the atmosphere as a result of human activity.

Alkali metal

The Group 1 elements.

Alkaline

Having a pH greater than 7.

Allotropes

Different forms of the same element in the same physical state.

Alloy

A mixture containing mostly metals.

Allred and Rochow electronegativity scale

An empirical scale of electronegativity.

Alpha decay

Nuclear decay by emission of an alpha particle.

Aaulgam

An alloy that contains mercury.

Amorphous solid

A noncrystalline solid with no well-defined ordered structure.

Amphoteric, ampholyte

A substance that can act as either an acid or a base in a reaction.

Anatase; titanium oxide

A mineral containing titanium (TiO_2).

Angstrom (Å); Ångstrom (units)
A non-SI unit of length.

Anion
A negatively charged ion such as the hydroxide ion OH$^-$.

Aqua regia
A strong acid containing one part concentrated nitric acid and three parts concentrated hydrochloric acid.

Aqueous solution
A solution in which water is the dissolving medium, or solvent.

Arrhenius equation
An equation expressing the dependence of the rate constant on the absolute temperature.

Asbestos
A naturally occurring fibrous mineral.

Aspirin (acetylsalicylic acid)
A drug in the family of salicylates.

Association colloid
A colloid in which the dispersed phase consists of micelles.

Atmosphere (air)
The mixture of gases (air) that surrounds the earth's surface.

Atomic mass unit (amu); dalton
One twelfth of the mass of the carbon-12 atom.

Atomic number (Z)
The number of protons in the nucleus of an atom; each element has a unique atomic number.

Atomic orbital
A wavefunction that describes the behavior of an electron in an atom.

Atomic radius
Half the distance between the atomic molecule consisting of identical atoms.

Auto-ionization
The transfer of a proton from one molecule to another of the same substance.

Avogadro constant (LA)
The number of carbon-12 atoms in 0.012 kg of carbon-12.

Avogadro's law
Equal volumes of gases at the same temperature and pressure contain the same number of particles (atoms or molecules).

Balance a chemical equation; manual balancing
The process of balancing a chemical equation involves finding the whole number coefficients which give the same amount of each element on each side of the equation.

bar

A unit of pressure.

Barometer

A device for measuring the pressure of the atmosphere.

Base

The most common definition, due to Arrhenius is that a substance that produces hydroxide ions, when it dissolves in water.

Base dissociation constant; base ionization constant

The equilibrium constant for the ionization of a weak base in water, giving the hydroxide ion and the conjugate acid of the base.

Benzene (C_6H_6)

Benzene is an aromatic hydrocarbon, C_6H_6 (CAS:71-43-2).

beta particle (β-)

An electron emitted by an unstable nucleus, when a neutron decays into a proton and an electron.

Binary compound

A compound composed of only two elements (for example, CuO).

Biochemistry

The chemistry of living things.

Bleach

A dilute solution of sodium hypochlorite or calcium hypochlorite which kills bacteria and destroys colored organic materials by oxidizing them.

Body-centered cubic unit cell

A cubic unit cell in which there is a lattice point at the center of the unit cell as well as at the comers.

Bohr atom; Bohr

A simple model of the atom (hydrogen).

Bohr radius; (ao) bohr

The atomic unit of length.

Boiling point

The temperature at which the vapor pressure of the liquid equals the atmospheric pressure.

Boltzmann constant (k)

A fundamental constant equal to the ideal gas law constant divided by Avogadro's number.

Bond energy

Energy change per mole when a bond is broken in the gas phase for a particular substance.

Bond enthalpy

Enthalpy change per mole when a bond is broken in the gas phase for a particular substance.

Bond length

The distance between atomic centers involved in a chemical bond.

Bond strength

A measure of how difficult it is to break a chemical bond.

Borax

A white, yellowish mineral containing boron with a chemical formula of $Na_2[B_4O_5(OH)_4] \cdot 8H_2O$. It has a Mohs hardness of 2 and a specific gravity of 1.7 -1.8.

Brass

A shiny yellow to yellow-orange alloy that contains about two parts copper for every one part zinc.

Brönsted base

A material that accepts hydrogen ions (proton) in a chemical reaction.

Brönsted-Lowry definition of acids and bases

An acid is a proton donor and a base is a proton acceptor in a proton-transfer reaction.

Bronze

A yellow to yellow-brown alloy that contains mostly copper and tin, with small amounts of other metals.

Brownian motion; Brownian movement

Small particles suspended in liquid move spontaneously in a random fashion.

Buckminsterfullerene; C_{60}; fullerene; buckyball

A form of carbon consisting of 60 carbon atoms bound together to make a roughly spherical "buckyball".

Buffer; pH buffer; buffer solution

A solution that resist a change in pH or can maintain its pH value with little change upon addition of small amounts of acids or bases.

Buffer capacity

The ability of a buffered solution to absorb protons or hydroxide ions without a significant change in pH.

Buffered solution

A solution that resists a change in its pH when either hydroxide ions or protons are added.

Bunsen burner

A gas burner with adjustable air intake, commonly used in laboratories.

Buret; burette

A cylindrical glass tube closed by a stopcock on one end and open on the other, with volume gradations marked on the barrel of the tube, used to precisely dispense a measured amount of a liquid.

Calorie (cal)

A non-SI unit of energy.

Calorimeter

An insulated vessel for measuring the amount of heat absorbed or released by a chemical or physical change.

Calorimetry

Experimental determination of heat absorbed or released by a chemical or physical change.

Candela (cd)

The SI base unit of luminous intensity.

Carat (Diamond)

A unit of measurement of the weight of diamond.

Carbanion

An organic anion in which carbon bears a negative charge and has an unshared pair of electrons.

Carbocation; carbonium ion

A carbocation is an ion with a positively-charged carbon atom.

Carbohydrate

A class of organic compounds including sugars and starches.

Carbon monoxide (CO)

A colorless, odorless gas resulting from the incomplete combustion of hydrocarbon fuels.

Carboxyl group

The—COOH group.

Catalyst; catalyze; catalysis

A substance that increases the rate of a chemical reaction, without being consumed or produced by the reaction.

Cathode

The electrode at which reduction occurs in a cell.

Cation

A positively charged ion such as the hydrogen ion H^+.

Cellulose

A polysaccharide.

Celsius (℃); Celsius temperature scale; Celsius scale

A common but non-SI unit of temperature.

CGPM (General Conference on Weights Measures)

The primary intergovernmental treaty organization responsible for the SI, representing nearly 50 countries.

Chemical equation

The symbolic representation of a in terms of chemical formulas.

Chemical bond

A strong attractive force that exists certain atoms in a substance.

Chemical equilibrium

The state of a dynamic reaction system in which the concentrations of all reactants and products remain constant as a function of time.

Chemical kinetics

A branch of physical chemistry which deals with the rates and mechanisms of chemical reactions.

Chemical oceanography

The study of the chemical composition of seawater.

Chemical stoichiometry

The calculation of the quantities of materials consumed and produced in a chemical reaction.

Chlorofluorocarbons (CFCs)

Compounds containing carbon, chlorine and fluorine.

Cis-trans isomerism

A type of geometrical isomerism.

CODATA

Committee on Data for Science and Technology.

Colligative property

A solution property that depends on the number of solute particles present.

Combustible material

Any material which will catch fire and bum.

Complexing agent; complexant

A ligand that binds to a metal ion to form a complex.

Complexometric titration; chelometric titration

A titration based on a reaction between a ligand and a metal ion to form a complex.

Composite

A material constructed of two or more different kinds of materials in separate phases.

Compound

A substance composed of atoms of two or more dements chemically combined in fixed proportions.

Compton effect

Demonstrates that photons have momentum.

Computer chemistry

Use of computers as a computational tool or for analysis of chemical data.

Condensation

The change of a gas to either the liquid or the solid state.

Condensation reaction

A reaction in which two molecules or ions are chemically combined by the elimination of a small molecule such as water.

Congener

Elements belonging to the same group on the periodic table.

Conjugate acid

In conjugate acid-base pair, conjugate acid is the species that can donate a proton.

Conjugate acid-base pair

Two species in an acid-base reaction, one acid and one base, that differ by the loss or gain of a proton.

Conjugate base

In conjugate acid-base pair, conjugate acid is the species that can accept a proton.

Corrosive material

One which causes damage to skin, eyes or other parts on the body on contact.

Control of substances hazardous to health (COSHH)

Regulations to promote safe working with potentially hazardous chemicals.

Covalent bond; covalent bonding

A type of bonding in which atoms share electrons.

Critical reaction

A reaction in which exactly one neutron from each fission event causes another fission event, thus sustaining the chain reaction.

Crystal structure

The exact arrangement of molecules or atoms in a crystal.

Crystallization

The process of forming a crystalline structure.

Crystals

Solid materials consisting of a regularly-repeated unit.

Curie (Ci)

A unit of radioactivity.

Cutaneous hazard

A chemical which may cause harm to the skin, such as defatting, irritation, skin rashes or dermatitis.

Dalton; atomic mass unit (amu)

One twelfth of the mass of the carbon-12 atom.

Debye(D)

A unit used to express dipole moments.

Decant

To draw off the upper layer of liquid after the heaviest material (a solid or another liquid) has settled.

Density

The mass per unit volume of a substance or solution.

Desalination

The removal of ions from salty water or seawater to make drinkable or industrially usable water.

Desiccant

A drying agent; a chemical agent that absorbs moisture.

Desiccator

An apparatus for drying substances by absorbing the moisture present in a chemical substance.

Detergent

A surfactant (or a mixture containing one or more surfactants) having cleaning properties in dilute solution (soaps are surfactants and detergents).

Deuteron

The ion $^2H^+$ or D^+ where D is the hydrogen isotope ^2H having a mass number of 2.

Diatomic molecule

A molecule composed of two atoms.

Diffusion; diffuse

The mixing of two substances caused by random molecular motions.

Dilatometer

A device for measuring volume change.

Dilution

The process of adding solvent to lower the concentration of solute in a solution.

Dimer

A molecule composed of two subunits linked together.

Dipole; electric dipole

An object whose centers of positive and negative charge do not coincide.

Dipole moment

A property of a molecule whereby the charge distribution can be represented by a center of positive charge and a center of negative charge.

Dipole-dipole attraction

The attractive force resulting when polar molecules line up such that the positive and negative ends are close to each other.

Dipole-dipole interaction

Intermolecular or intramolecular interaction between molecules or groups having a permanent electric dipole moment.

Distillation

A method for separating the components of a liquid mixture that depends on differences in the ease of vaporization of the components.

DNA

Deoxyribonucleic acid, the basic building-block of life.

Double bond

A bond in which two atoms share two pairs of electrons.

Dumas method

A method used to determine the molar mass (molecular weight) of volatile liquids.

Effusion

The process by which a gas, under pressure, escapes from a vessel by passing through a very small opening.

Electric charge

A fundamental property of elementary particles.

Electric dipole

An object whose centers of positive and negative charge do not coincide.

Electric dipole moment (g); dipole moment

A measure of the degree of polarity of a polar molecule.

Electric field

A field of forces that act on any electric charge placed within it.

Electromagnetic radiation

Oscillating electric and magnetic fields moving through a medium perpendicular to each other.

Electromotive force (or emf) *(E)*

Measured in volts, defined as the rate at which work is done electrically on a circuit divided by the current.

Electron (e^-)

A fundamental constituent of matter, having a negative charge.

Electron affinity

The energy change for the process of adding an electron to a neutral atom in the gaseous state to form a negative ion.

Electron configuration of atoms

A list showing how many electrons are in each orbital or subshell.

Electron microscope

A microscope that uses a source of electrons to attain high magnification levels.

Electron volt (eV)

A unit of energy.

Electronegative

An atom in a molecule which attracts electrons towards it.

Electronegativity

The tendency of an atom in a molecule to attract electrons toward it.

Electronic transition

The transfer of an electron from one electronic energy level to another in an atom, a molecule or an ion.

Electrostatic forces

The forces which exist between particles which are electrically charged.

Empirical formula

The formula of a substance written with the smallest whole-number subscripts.

Empirical law

A law strictly based on experiment, which may lack theoretical foundation.

Equimolar

Containing equal number of moles.

Equimolecular

Containing equal number of molecules.

Equivalent point; end point

The point in titration when a stoichiometric amount of reactant has been delivered.

Erlenmeyer flask; conical flask

A flask having a wide base, narrow neck and conical form convenient in laboratory experimental work involving swirling liquids by hand; also called a conical flask.

Ethylenediaminetetracetic acid (EDTA)

A polydentate ligand that tightly complexes certain metal ions.

Evaporation

The conversion of a liquid into a vapor or gas.

Extensive property

A physical property whose magnitude depends upon the amount of material in a sample.

Filtration

The removal or separation of solid from a suspension.

Flammable limits

The concentration range under which a mixture of a flammable material (fuel) and air (oxygen) may produce a fire or explosion when an ignition source (such as a spark or open flame) is present.

Flammable material; inflammable

Very easy to ignite.

Flammable range

The difference between the lower and upper flammable limits.

Flash point

The temperature at which a liquid will yield enough flammable vapor to ignite.

Flocculation

The process in which small particles aggregate together to give bigger particles.

Florence flask

A round bottle having a flat bottom and a long neck for use in chemistry laboratories.

Forensic chemistry

The application of chemical scientific knowledge to questions of civil and criminal law.

Formal charge (FC)

A hypothetical charge defined as: FC = number of valence electrons of the atom - number of Lone pair electrons on this atom, half the total number of electrons participating in covalent bonds with this atom.

Fossil fuel

A fuel that consists of carbon-based molecules derived from decomposition of once-living organisms; coal, petroleum, or natural gas.

Frasch process; sulfur mining

A mining process in which the underground deposits of solid sulfur are melted in place by pumping in superheated steam and the molten sulfur is forced upward using compressed air.

Free radical

A molecule with an odd number of electrons.

Freeze-drying

A process in which a material is dried in a frozen state under high vacuum.

Frequency

Cycles per second.

Fullerene

A compound composed solely of an even number of carbon atoms which form a cage-like fused-ring polycyclic system.

Functional group

Part of a molecule with a characteristic reaction or property.

Gangue

The worthless portion of an ore.

Geiger counter

A radiation detector.

Gene

A given segment of the DNA molecule that contains the code for a specific protein.

Geochemistry; geological chemistry

The study of materials and chemical reactions in rocks, minerals, magma, seawater, and soil.

Graphite

An amorphous form of carbon, made of carbon atoms bound hexagonally in sheets.

Group (of the periodic table)

A vertical column of elements having the same valence electron configuration and similar chemical properties.

Guy-Lussac's law; Charles's law

The volume of a fixed amount of gas, at constant pressure, increases linearly with the temperature

Haber process

The manufacture of ammonia from nitrogen and hydrogen.

Half-life (of a radioactive sample)

The time required for the number of nuclides in a radioactive sample to reach half the original number of nuclides.

Half-reactions

The two parts of an oxidation-reduction reaction, one representing oxidation, the other reduction.

Halon

A compound consisting of bromine, fluorine, and carbon.

Hard soft acid base concept; Pearson HSAB rule

A qualitative concept originated in 1963 by Ralph G. Pearson to describe the reactivities and stabilities of Lewis acids and bases according to their relative polarizabilities.

Hard water

Water from natural sources that contains relatively large concentrations of calcium and magnesium ions.

Heat

Energy transferred between two objects because of a temperature difference between them.

Heat capacity *(C)*

The amount of heat required to raise the temperature of an object by 1 degree Celsius.

Heat of atomization

The negative value of the heat of formation. It is equal to the total bond energy of a compound.

Henderson-Hasselbalch equation

An equation relating the pH of a buffer solution for different concentrations of conjugate acid and base.

Henry's law

An equation relating the solubility of a gas in a liquid.

Hertz (Hz)

Cycles per second.

High-spin complex ion

A complex ion in which there is minimum pairing of electrons in the orbitals of the metal ion.

Homologue

A compound belonging to a series of compounds differing from each other by a repeating unit.

Humidity

The moisture content of air.

Hydrate

A compound that contains water molecules weekly bound in its crystals.

Hydration

The interaction between solute particles and water molecules.

Hydrocarbons

Chemicals containing only carbon and hydrogen.

Hydrofluoric acid (HF)

A clear, corrosive liquid that has an extremely pungent odor and forms dense white vapor clouds if released.

Hydrogen peroxide (H_2O_2)

A reactive oxidant (H_2O_2) .

Hydrogen sulfide (H_2S)

A colorless, flammable, poisonous compound (gas) having a characteristic rotten-egg odor.

Hydron

General name for the ion H^+.

Hydronium ion

H_3O^+.

Hydrophilicity; hydrophilic

The tendency of a molecule to be solvated by water.

Hydrophobic; hydrophobic colloid

Lack of attraction towards water.

Hydrophobicity

The association of non-polar groups or molecules in an aqueous environment.

Hygroscopic

The tendency of a substance to absorb water from the atmosphere.

Hypothesis

One or more assumptions put forth to explain observed phenomena.

Ideal gas

A hypothetical gas that exactly obeys the ideal gas law. A real gas approaches ideal behavior at high temperature and/or low pressure.

Ideal gas law; ideal gas equation

An equation relating the properties of an ideal gas, expressed as $pV = nRT$.

Indigo

A dark blue crystalline powder which is a commercial dye.

Ion

An atom or a group of atoms that has a net positive or negative charge.

Ion-product constant (K_w)

The equilibrium constant for the auto-ionization of water; $K_w = c(H^+) \cdot c(OH^-)$.

Ionic bonding

The attraction between oppositely charged ions.

Ionic compound

A compound that results when a metal reacts with a nonmetal to form cations and anions.

Ionic equation

A chemical equation in which strong electrolytes are written as dissociated ions.

Ionic solid

A solid containing cations and anions that dissolves in water to give a solution containing the separated ions, which are mobile and thus free to conduct an electric current.

Ionization energy

The quantity of energy required to remove an electron from a gaseous atom or ion.

Isobaric

Having constant pressure.

Isochoric

Having constant volume.

Isoelectronic

A group of atoms or ions having the same number of electrons.

Isomer; structural isomer

Molecules with identical molecular formulas but different structural formulas.

Isomerization

A chemical change that involves a rearrangement of atoms and bonds within a molecule, without changing the molecular formula.

Isosteric

Having identical valence electron configurations.

Isothermal

Having constant temperature.

Isotone

One of a group of atoms or ions with nuclei that contain the same number of neutrons but different numbers of protons.

Isotonic; isotonic solution

Refers to solutions that have equal solute concentrations.

Isotope; isotopic; isotopy

Atoms or ions of an element with different numbers of neutrons in their atomic nucleus.

Isotopic abundance

The fraction of atoms of a given isotope in a sample of an element.

Isotopic mass

The mass of a single atom of a given isotope, usually given in atomic mass units (daltons).

IUPAC (International Union of Pure and Applied Chemistry)

An organization which sets international standards for chemical nomenclature.

Joule (J)

The SI unit of energy.

Kekulé structure (aromatic compounds)

A structural representation showing alternating single and double bonds of an aromatic molecular entity.

Kelvin (K)

The SI base unit of thermodynamic temperature.

Ketone

An organic compound containing the carbonyl group bonded to two carbon atoms.

Lanthanide

Elements 57-71.

Lanthanide contraction

An effect that causes sixth period elements with filled 4f subshells to be smaller than otherwise expected.

Lattice

A regular array of ions or atoms.

Law of conservation of energy

Energy can be converted from one form to another but can be neither created nor destroyed.

Law of conservation of mass

Mass is neither created nor destroyed.

Law of constant composition

A given compound always contains elements in exactly the same proportion by mass.

Law of mass action

A general description of the equilibrium condition; it defines the equilibrium expression.

Law of multiple proportions

A law stating that when two elements form a series of compounds, the ratios of the masses of the second element that combine with one gram of the first element can always be reduced to small whole numbers.

Le Chatelier's principle

When a stress is applied to an equilibrium mixture, the equilibrium will shift to relieve the stress.

Lead storage battery

A battery (used in cars) in which the anode is lead, the cathode is lead coated with lead dioxide, and the electrolyte is a sulfuric acid solution.

Lewis acid

A species that can form a covalent bond by accepting an electron pair from another species.

Lewis adduct

The adduct formed between a Lewis acid and a Lewis base.

Lewis base

A species that can form a covalent bond by donating an electron pair to another species.

Lewis structure; electron dot structure; dot structure

A model that represents the electronic structure of a molecule.

Ligand

A Lewis base (an atom, ion or functional group) that binds to a metal ion to form a complex.

Limestone

A sedimentary rock composed of the mineral calcite (calcium carbonate).

Limiting reactant; limiting reagent

The reactant that limits the amount of product produced in a chemical reaction.

Lipophilic; lipophilicity

Refers to a substance's solubility in fat.

Liquefaction

Changing a solid into a liquid.

Litmus

Pigments extracted from certain lichens that turns blue in basic solution and red in acidic solution.

Lone pair

An electron pair that is localized on a given atom; an electron pair not involved in bonding.

Lyman series

Tile series which describes the emission spectrum of hydrogen when electrons are jumping to the ground state. All of the lines are in the ultraviolet.

Macro-

A prefix meaning "large".

Main group elements

Elements of the s and p blocks.

Malleable; malleability

Cable of being hammered into sheets. Metals are typically malleable materials.

Manometer

An instrument for measuring gas pressures.

Mass number (A)

The total number of protons and neutrons in the nucleus of an element.

Medicinal chemistry

A branch of chemistry concerned with the discovery, design, synthesis, and investigation of biologically active compounds and reactions that these compounds undergo in living things.

Meniscus; meniscuses; menisci

A phase boundary that is curved because of surface tension.

Mercurous chloride; calomel

Mercuric monochloride Hg_2Cl_2.

Metabolism; metabolic; metabolic reaction

A sequence of biochemical reactions that converts fuel molecules into energy used to drive other biological processes.

Metabolite

A compound produced by metabolic reactions.

Metal; metallic

A substance that conducts heat and electricity, is shiny and reflects many colors of light, and can be hammered into sheets or drawn into wire.

Metal complexes

A structure composed of a central metal atom or ion, generally a cation, surrounded by a number of negatively charged ions or neutral molecules possessing lone pairs.

Metallic compounds

Compounds that contain at least one metallic element.

Metalloid; semimetal; semi-metal

An element having both metallic and nonmetallic properties.

Metallurgy

The process of separating a metal from its ore and preparing it for use.

Micelle

A colloidal-sized particle formed in water by the association of molecules.

Micro-

A prefix meaning "small".

Millimeters of mercury (mm Hg)

A unit of measurement for pressure.

Mineral

A naturally-occurring chemical compound.

Mineralogy

The study of minerals.

Mohs scale of mineral hardness

Characterizes the scratch resistance of various minerals.

Molarity; molar concentration

The number of moles of solute dissolved in one liter of solution. For example, a 2 mol/L solution of hydrochloric acid contains two moles of HCl per liter of solution.

Molar

Pertaining to moles.

Molar mass

The mass of one mole of a material.

Molar volume

The volume occupied by one mole of a material.

Molality; mdar concentration

Number of moles of solute dissolved in one kilogram of solvent.

Mole (mol)

The SI base unit of amount of substance.

Mole fraction

Moles of the substance per mole of mixture.

Molecular matt; relative molecular weight

The average mass of a molecule, calculated by summing the atomic weights of atoms in the molecular formula.

Molecular sieve

A material that contains many small cavities interconnected with pores of precisely uniform size.

Molecular structure

The three-dimensional arrangement of atoms in a molecule.

Molecule

The smallest division of a compound that still retains or exhibits all the properties of the substance.

Monomer

Compounds used to make a polymer.

Monoprotic add

An acid with one acidic proton.

Monosaccharides

Carbohydrates in the form of simple sugars, composed of one saccharide unit.

Nanotechnology; nanomaterials

Involves the manipulation of matter at nanometer length (one-billionth of a meter) scales to produce new materials, structures and devices.

Natural abundance

Percentage of an element occurring on earth in a particular stable isotopic form.

Natural gas

Consists of mostly methane and is associated with petroleum deposits.

Natural law

A statement that expresses generally observed behavior.

Net ionic equation

An equation for a reaction in solution, representing strong electrolytes as ions and showing only those components that are directly involved in the chemical change.

Network solid

An atomic solid containing strong directional covalent bonds.

Neutralization reaction

An acid-base reaction.

Neutron

A subatomic particle with no net electric charge.

Noble gases; inert gases; rare gases

The Group VIIIA elements which are all monoatomic gases.

Nomenclature

A system for naming things.

Nonelectrolyte

A substance that, when dissolved in water, gives a nonconducting solution.

Nonmetal

A substance that conducts heat and electricity poorly.

Nonpolar

Having a relatively even or symmetrical distribution of charge.

Nonpolar molecule

A molecule in which the center of positive charge and the center of negative charge coincide.

Normal boiling point

The boiling temperature under one atmosphere of pressure.

Normal melting/freezing point

The melting/freezing point of a solid at a total pressure of one atmosphere.

Nuclear atom

The modem concept of the atom as having a dense center of positive charge (the nucleus) and electrons moving around the outside.

Nuclear fission

The splitting of a nucleus into two smaller nuclei and neutrons.

Nuclear fusion

The combination of two smaller nuclei to form a larger nucleus.

Nucleon

A particle in an atomic nucleus, either a neutron or a proton.

Nucleus

The small, dense center of positive charge in an atom.

Nuclide

The general term applied to each unique atom.

Nylon

A synthetic condensation polymer made of S repeating units with amide linkages.

Octahedral geometry

The geometry of a molecule in which six atoms s occupy the vertices of a regular octahedron.

Octahedron

A figure with eight faces and six vertices.

Octet rule

The tendency of atoms in molecules to have eight electrons in their valence shells (two for hydrogen atom).

Oligosaccharides

A carbohydrate composed of several saccharide units (usually 3-10).

Orbital

A representation of the space occupied by an electron in an atom (or molecule).

Ore

A mineral, deposit or rock from which a metal or nonmetal can be extracted.

Ostwald process

An industrial preparation of nitric acid starting from the catalytic oxidation of ammonia to nitric oxide.

Oxidation; oxidize; oxidizing; oxidized

The loss of one or more electrons by an atom, molecule, or ion.

Oxidation number; oxidation state; positive valence

A convention for representing a charge of an atom embedded within a compound.

Oxidation reduction reaction; redox reaction

A reaction in which electrons are transferred between species or in which atoms change oxidation number.

Oxidation state

A concept that provides a way to keep track of electrons in oxidation-reduction reactions according to certain rules.

Oxide

A binary compound that contains oxygen in the -2 oxidation state.

Oxidizing agent; oxidant; oxidizer

A reactant that removing electrons from other reactants in a chemical reaction.

Oxyacid

An acid in which the acidic proton is attached to an oxygen atom.

Ozone layer

The region of the stratosphere containing the bulk of atmospheric ozone.

Ozone (O_3)

A gas composed of three atoms of oxygen (O_3).

Ozone-depleting substance (ODS)

A substance that contributes to stratospheric ozone depletion.

Pairing energy (Crystal field theory)

The energy required to put two electrons into the same orbital.

Partial miscibility; partially miscible

Two liquids are considered partially miscible if shaking equal volumes of the liquids together results in a meniscus visible between two layers of liquid.

Parts per million (ppm)

Parts of solute per million parts of solution.

Pascal (Pa)

The SI unit of pressure.

Pauling Electronegativity Scale

An empirical scale of electronegativity.

Percent yield

The actual yield of a product as a percentage of the theoretical yield.

Percentage composition

The mass percentages of each element in a compound.

Period (periodic table)

A row of elements in the periodic table.

Periodic law

Physical and chemical properties of the elements recur in a regular way when the elements are arranged in order of increasing atomic number.

Periodic table

An arrangement of the elements according to increasing atomic number that shows relationships between element properties.

Periodic trend

A regular variation in element propertied with increasing atomic number that is ultimately due to regular variations in atomic structure.

Peroxide

A compound with oxygen in the -1 oxidation state such as hydrogen peroxide.

Petroleum

A thick, dark liquid composed mostly of hydrocarbon compounds.

pH

An acidity scale defined as: $pH = -\lg c(H^+)$.

Phase

An entity of a material system which is uniform in chemical composition.

Phase diagram

A diagram that shows the conditions under which the different states of a substance are stable.

Phenolphthalein

An organic compound used as an acid-base indicator.

Photochemistry

A branch of physical chemistry, dealing with the study of chemical changes caused by light.

Photon

A particle of electromagnetic radiation.

pK_a

$pK_a = -\lg K_a$

pK_b

$pK_b = -\lg K_b$

Plasma

An electrically neutral gas of ions and electrons.

Polar covalent bond

A covalent bond with non-uniform distribution of electrons.

Polar molecule

A molecule that has a permanent dipole moment.

Polyatomic ion

An ion containing a number of atoms.

Polymer

A large, usually chain-like molecule of very high molar mass that is made up from many repeating units (monomers) of low molar mass.

Polymerization

A process in which many small molecules (monomers) are joined together to form a large molecule.

polymorphism

The ability of one molecule to crystallize into more than one crystal structure.

Polyprotic acid

An acid with more than one acidic proton.

Ponds per square inch (psi)

Pounds per square inch (lb/in^2) or psi is a unit of pressure; 1 psi = 51.7 mmHg = 6890 Pa.

Precipitate

An insoluble solid compound formed during a chemical reaction in solution.

Precipitation reaction

A reaction in which an insoluble substance forms and separates from the solution as a solid.

Pressure (p)

Force per unit area.

Proton

A fundamental constituent of matter, having a positive charge.

Pure substance

A substance with constant composition.

Quartz

A mineral consisting of silicon dioxide.

Reactant

A starting substance in a chemical reaction.

Reaction intermediate

A reactive species produced during a chemical reaction.

Reamer temperature scale

A temperature scale in which, under a pressure of 1 atmosphere, the ice point is 0 degrees and the boiling point of water is 80 degrees.

Reducing agent (electron donor)

A reactant that donates electrons to another substance.

Reduction

A decrease in oxidation state.

Representative elements

Same as Main group elements.

Resonance (Lewis structures)

A condition occurring when more than one valid Lewis structure can be written for a particular molecule.

Reverse osmosis (RO)

A process in which a solvent, such as water, is forced by a pressure greater than the osmotic pressure to flow through a semipermeable membrane from the concentrated solution to a more dilute one.

Roentgen equivalent man (rem)

A unit of radiation dosage.

Salt

An ionic compound.

Saturated solution

A solution that contains as much solute as can be dissolved in that solution.

Scientific method

A process of studying natural phenomena.

second (s)

The SI base unit of time.

Secondary structure (of a protein)

The three-dimensional structure of the protein chain.

SI units

The International System of Units.

Sigma bond

A covalent bond.

Silica

Silicon-oxygen compound.

Silica gel

A porous, granular form of silica, synthetically manufactured from sodium silicate.

Silicates

Salts that contain metal cations and polyatomic silicon - oxygen anions that are usually polymeric.

Simplest formula; empirical formula

The simplest whole-number ratio of atoms in a compound.

Single bond

A bond in which two atoms share one pair of electrons.

Sol

A colloid with solid particles suspended in a liquid.

Solid

One of the three states of matter.

Solubility

The amount of a substance that dissolves in a given volume of solvent.

Solubility product

The constant for the equilibrium expression representing the dissolving of an ionic solid in water.

Solute

A substance dissolved in a solvent to form a solution.

Solution

A homogeneous mixture.

Solvay process

An industrial process for obtaining sodium carbonate from sodium chloride and limestone.

Solvent

The dissolving medium in a solution.

Specific gravity; relative density

A measure of the density of a material.

Spectator ions

Ions present in solution that do not participate directly in a reaction.

Standard atmosphere (atm)

A unit of pressure.

Starch

A carbohydrate (polysaccharide) insoluble in water.

States of matter

The three different forms in which matter can exist: solid, liquid, and gas.

Stoichiometric coefficient

The coefficients given before substances in a balanced chemical equation.

Stoichiometry

The relationship between the amounts of substances in a balanced chemical reaction.

STP (Standard temperature and pressure)

The condition 0 ℃ (273.15 K) and 1 atm pressure.

Strengths of Acids and Bases

Approximate molarities and specific gravities for some acids and bases.

Structural formula

A chemical formula that shows how the atoms are bonded to one another in a molecule.

Structural isomers

Compounds that contain the same number of each atom (same chemical formula) but differs in how the atoms are joined together.

Sublimation

The change of a solid directly to the gaseous state.

Substituent

An atom or group of bonded atoms that can be considered to have replaced a hydrogen atom (or two hydrogen atoms in the special case of bivalent groups) in a parent molecular entity (real or hypothetical).

Substrate

A reactant molecule.

Superacid

A medium having a high acidity.

Surfactant; surface active agent

A substance which lowers the surface tension of the medium in which it is dissolved.

Surroundings

Everything in the universe surrounding a thermodynamic system.

System (Thermodynamics)

That part of the universe on which attention is to be focused.

Tertiary structure (of a protein)

The overall shape of a protein.

Theoretical yield

The maximum amount of a given product that can be formed when the limiting reactant is completely consumed.

torr

A non-SI unit of pressure equivalent to 1 mm Hg.

Toxicology

A scientific discipline involving the study of chemistry, biochemistry.

trans

A Latin word meaning "on the opposite side".

Transuranium elements

The elements beyond uranium.

Triple bond

A bond in which two atoms share three pairs of electrons.

Triton

The ion $^3H^+$ or T^+ where T is the hydrogen isotope ^3H having a mass number of 3.

Unified atomic mass unit (u)

One twelfth of the mass of the carbon-12 atom.

Universal gas constant (R)

The combined proportionality constant in the ideal gas law.

Unsaturated solution

A solution in which more solute can be dissolved than is dissolved already.

Vacuum

A region of space where no matter is present.

Valence bond (VB) theory

A method of approximating the total wavefunction of a molecule.

Valence electrons

The electrons in the outermost occupied principal quantum level of an atom.

Valence shell electron pair repulsion model (VSEPR)

A simple model of bonding.

van der Waals (VDW) interactions

Noncovalent interactions.

Vapor

The gaseous state of any substance that normally exists as a liquid or solid.

Vapor pressure

The pressure of the vapor over a liquid at equilibrium in a closed container.

Volatile

Refers to a liquid or solid having a relatively high vapor pressure at normal temperature.

Wavelength

The distance between two consecutive peaks or troughs in a wave.

Weak acid

An acid that dissociates only to a slight extent in aqueous solution.

Weak base

A base which dissociates partially when dissolved in water.

Weak electrolyte

A material that, when dissolved in water, gives a solution that conducts only a small electric current.

Zincography

The process of etching unprotected parts of a zinc plate with strong acids to produce a printing surface.

Zwitterion (or double ion)

An ion carrying both a positive and negative charge.

Appendix 8: Terms Used in Organic and Biological Chemistry

Acetal

A molecule with two single bonded oxygens attached to the same carbon atom.

Acetate ($CH_3COO\sim$, $C_2H_3O_2\sim$); acetate ion

Ion or group derived from acetic acid.

Acetic acid; ethanoic acid; vinegar

A simple organic carboxylic acid that gives vinegar its characteristic odor and flavor.

Acetin; glyceryl monoacetate

An organic ester produced by the partial esterification of glycerol.

Acetoin; acetylmethylcarbinol; dimethylketol

An organic compound used chiefly in the manufacture of flavors and essences.

Acetone; dimethylketone; 2-propanone

An organic carbonyl compound (ketone).

Acetonitrile; methyl cyanide

An organic solvent (nitrile), also called methyl cyanide CH_3CN.

Acetophenone; phenyl methyl ketone

An aromatic carbonyl compound ($C_6H_5COCH_3$).

Acetostearin

A waxlike nongreasy solid, $C_{23}H_{44}O_5$, used chiefly as a food preservative and as a plasticizer.

Acetyl chloride

An organic acid chloride (CH_3COCl).

Acetyl choline

The acetic acid ester of choline ($C_7H_{17}NO_3$).

Acetyl coenzyme A; acetyl CoA

The acetylated form of coenzyme A, formed as an intermediate in the oxidation of carbohydrates, proteins and fats in animal metabolism.

Acetylcysteine

An organic compound, $C_5H_9NO_3S$, used in solution as an inhalant to dissolve mucus in the treatment of chronic bronchitis.

Acetylecholinesterase enzyme

An enzyme that hydrolyze the neurotransmitter acetylcholine.

Achiral molecule

A molecule with no chiral center.

Acriflavine; euflavine; tryptaflavine

An orange brown granular solid, $C_{14}H_{14}N_3CL$ used chiefly as an antiseptic in medicine.

Acrylate

A salt or an ester of acrylic acid.

Acrylic acid

A colorless, corrosive liquid, $C_3H_4O_2$ having a strong acrid odor used as a monomer in the polymer industry.

Alcohol

A class of organic compounds composed of carbon, hydrogen and oxygen, with—OH group.

Aldehyde

An organic compound containing the carbonyl group bonded to at least one hydrogen atom.

Aromatic

A type of hydrocarbon, such as benzene or toluene. Some aromatics are toxic.

Aromaticity; aramatic; antiaromaticity; antiaromatic

Having a chemistry typified by benzene.

Automerization; degenerate rearrangement; permutetional isomerization; topomerization

A molecular rearrangement in which the reactant is transformed to the product which differs from reactant only in the permutation of identical atoms.

Bathochromic shift

Shift of a spectral band to lower frequencies (longer wavelengths) owing to the influence of substitution or a change in environment; red-shift.

Berry pseudorotation

A mechanism for the interconversion of trigonal bipyramid structures through an intermediate (or transition state) tetragonal pyramid structure.

Biradical

An even-electron molecular entity with two (in some cases delocalized) radical centers which act almost independently of each other.

Biradicaloid

A biradical displaying a strong coupling between the radical centers.

Carbohydrate

A class of organic compounds including sugars and starches.

Carrier-linked prodrug

A prodrug that contains a temporary linkage.

Chiral; chirality

Having nonsuperimposable mirror images. For example, a shoe or a glove is chiral.

Chiral center; asymmetric center

An atom in a molecule that causes chirality, usually an atom that is bound to four different groups.

***cis; trans* (Stereochemistry)**

Cis is a Latin word meaning "on the same side"; *trans* is a Latin word meaning "on the opposite side".

Combinatorial chemistry

The generation of large collections, or "libraries", of molecules by synthesizing all possible combinations of a set of smaller chemical structures.

Computational chemistry

A branch of chemistry concerned with the prediction or simulation of chemical properties, structures, or processes using numerical computational techniques.

Conjugated system

A system of atoms covalently bonded with alternating single and multiple bonds.

Conservation of orbital symmetry

The orbital symmetry control of concerted reactions.

Copolymer

A material produced by polymerizing a mixture of two (or more) different monomers.

Coulomb integral (Hückel)

A quantum mechanical integral between atomic orbitals treated as an empirical parameter.

Craig plot

A plot of two substituent parameters.

Degenerate rearrangement

See Automerization.

Dextrorotatory; dextro; D-

An optically active substance that rotates the plane of plane polarized light clockwise.

Diglyme

An abbreviation for a chemical substance, diethyleneglycol dimethylether.

Dimroth-Reichardt ET parameter

A measure of the ionizing power of a solvent.

Disaccharide

A sugar (carbohydrate) formed from two monosaccharides joined by a glycoside linkage.

Donor number (DN); GUTMANN

A quantitative measure of Lewis basicity devised by GUTMANN (1976).

Electronic configuration

The allocation of electrons within an atom or a molecule to a set of correspondingly atomic or molecular orbitals complying with the Pauli exclusion principle.

Electronic state

An arrangement allowed by the laws of quantum mechanics of electrons within an atom, molecule (or system of molecules).

Enantiomer

One of the two non-superimposable mirror-image forms of an optically active molecule.

Equilibrium geometry

Molecular geometry that corresponds to the true minimum on the respective potential energy surface.

Fat (glyceride)

An ester composed of glycerol and fatty acids.

Fatty acid

A long-chain carboxylic acid.

Fluxional molecules

A subclass of structurally nonrigid molecules in which all the interconverting species that are observable are chemically and structurally equivalent.

Force field (Molecular mechanics)

A set of potential functions defining bond stretch, bond angle (both valence and dihedral) distortion energy of a molecule as compared with its nonstrained conformation.

Free valence

A reactivity index applied mostly to radical reactions of conjugated systems.

Frontier orbitals

A HOMO and a LUMO of a molecule.

Hückel(4n + 2) rule; Hückel

A simple rule to predict if a cyclic conjugated molecule is aromatic based on the number of pi electrons.

Hammett equation; Hammett relation

An equation relating the logarithm of the relative rate constants or equilibrium constants of a reaction to substituent parameters.

Hammond principle; Hammond postulate

The hypothesis that, when a transition state leading to an unstable reaction intermediate has nearly the same energy as that intermediate, the two are interconverted with only a small reorganization of molecular structure.

Heterolytic fission; heterolytic bond cleavage

The process in which a covalent bond is broken with one atom taking both electrons forming the bond.

Humic matter

Decomposed organic matter found in soil.

Hydrocarbons

Chemicals containing only carbon and hydrogen.

Hyperconjugation

A model to account for the interaction of C—H and C—C sigma bonds with the conjugated pi-system.

Hypsochromic shift

Shift of a spectral band to higher frequency or shorter wavelength upon substitution or change in medium.

Identity reaction

A chemical reaction whose products are chemically identical with the reactants, for example the bimolecular self exchange reaction of CH_3I with I^-.

Inductive effect

An experimentally observable effect of the transmission of charge through a chain of atoms in a molecule by electrostatic induction.

Isogyric reaction

A reaction in which the number of electron pairs in reactants and products is conserved.

Isotopic fractionation factor

The ratio of isotopic abundances in molecules undergoing isotope exchange reaction.

Isotopologue

A molecular entity that differs only in isotopic composition (number of isotopic substitutions), e.g., CH_4, CH_3D, CH_2D_2.

Isotopomer

Isomers having the same number of each isotopic atom but differing in their positions. The term is a contraction of "isotopic isomer".

Lactose

A disaccharide (white powder) with a sweet taste. Lactose is also called milk sugar.

Levorotatory; levo: L-

An optically active substance that rotates the plane of plane polarized light counterclockwise.

Lewis acidity

The thermodynamic tendency of a substrate to act as a Lewis acid.

Lewis basicity

The thermodynamic tendency of a substance to act as a Lewis base.

Lipid

A diverse group of organic molecules that contain long hydrocarbon chains or rings and are hydrophobic.

Möbius aromaticity; Möbius

Predicting the aromatic character of a cyclic conjugated system.

Menthol

Menthol, $C_{10}H_{19}OH$, is an organic compound made synthetically or obtained from peppermint or other mint oils.

Meso compound

Or meso isomer, a non-optically active member of a set of stereoisomers, at least two of which are optically active.

Mesomeric effect

The intramolecular polarization of conjugated molecular system brought about by a substituent whose p_π or π-orbitals overlap with the π-MOs of the conjugated moiety.

Mesomerism

The term is particularly associated with the picture of π electrons as less localized in an actual molecule than in a Lewis formula.

Methacrylic acid; methyl methacrylate (MMA)

Methacrylic acid, $CH_2 = C(CH_3)COOH$, a colorless highly corrosive liquid with a penetrating odor.

Methyl acrylate; acrylic ester

An ester of acrylic acid.

Moiety

Part of a molecule.

Molecular entity

Any constitutionally or isotopically distinct atom, molecule, ion, ion pair, radical, radical ion, complex, conformer etc., identifiable as a separately distinguishable entity.

Molecular mechanics

Method of calculation of geometrical and energy characteristics of molecular entities on the basis of empirical potential functions (force field) the form of which is taken from classical mechanics.

Nucleophile

A nucleophile (literally nucleus lover as in nucleus) is a reagent that forms a chemical bond to its reaction partner (the electrophile) by donating both bonding electrons.

Nucleophilic

Able to approach the nuclei by the affinity of a nucleophile.

Olefin

A class of unsaturated hydrocarbons having the general formula C_nH_{2n} such as ethylene (C_2H_4, $H_2C = CH_2$).

Oligomer

A molecule composed of few subunits (monomers) linked together.

Pericyclic reaction

A chemical reaction proceeding through a folly conjugated cyclic transition state.

Permutational isomerization

See Automerization.

Phospholipid

An ester of glycerol with two fatty acids and phosphoric acid.

Phytochemistry; phytochemical

The study of substances found in plants. Phytochemicals are materials extracted from plant tissue.

Polynuclear aromatic hydrocarbons (PAHs)

Large ring aromatic hydrocarbons.

Protein

A natural polymer formed by condensation reactions between amino acids.

Pseudorotation

A conformational change resulting in a structure that appears to have been produced by rotation of the entire initial molecule and is superimposable on the initial one, unless different positions are distinguished by substitution or isotopic labeling.

Pyrene

A polycyclic aromatic crystalline hydrocarbon, $C_{16}H_{10}$ consisting of four fused benzene rings, found in coal tar and believed to be carcinogenic.

Racemic mixture

Or racemate, is one that has equal amounts of left- and right-handed enantiomers of a chiral molecule.

Resonance effect

See Mesomeric effect.

SOMO

An acronym for a singly occupied molecular orbital (such as the half-filled HOMO of a radical).

Spin crossover

A type of molecular magnetism that is the result of electronic instability caused by external constraints (temperature, pressure, or electromagnetic radiation).

Spin polarization

Static and dynamic spin polarization effects are distinguished.

Stereochemical nonrigidity

The capability of a molecule to undergo fast and reversible intramolecular isomerization.

Steric effect; steric hindrance

The effect on a chemical or physical property (structure, rate or equilibrium constant) upon introduction of substituents having different steric requirements.

Steric isotope effect

A secondary isotope effect attributed to the different vibrational amplitudes of isotopologues.

Strain energy; steric energy

The excess energy due to steric strain of a molecular entity or transition state structure.

Structure-based design

A design strategy for new chemical entities based on the three-dimensional (3D) structure of the target.

Substituent

An atom or group of bonded atoms that can be considered to have replaced a hydrogen atom (or two hydrogen atoms in the special case of bivalent groups) in a parent molecular entity (real or hypothetical).

Supermolecule

A discrete oligomolecular species that results from the intermolecular association of its components.

Supramolecular chemistry

A field of chemistry related to species of greater complexity than molecules, that are held together and organized by means of intermolecular interactions.

***syn; anti* (stereochemistry)**

Terms used to designate stereoisomers or stereochemical relationships among substituents.

Synchronization (principle of nonperfect synchronization)

Applies to reactions in which there is a lack of synchronization between bond formation or bond rupture and other primitive changes that affect the stability of products and reactants.

Synchronous concerted process

A concerted process in which the primitive changes concerned (generally bond rupture and bond formation) have progressed to the same extent at the transition state is said to be synchronous.

Taft steric parameter *(E_s)*

A relative reaction parameter encoding the reaction rate retardation due to the size of a substituent group.

Thermodynamic (equilibrium) isotope effect

The effect of isotopic substitution on an equilibrium constant.

Through-bond interaction

An intramolecular orbital interaction of spatially separated orbitals, where the orbitals interact through their mutual mixing with sigma-orbitals of the intervening framework.

Through-space interaction

An orbital interaction that results from direct spatial overlap of two orbitals.

Topomerization

See Automerization.

United atom approach

A simplification used by molecular mechanics programs such as AMBER and CHARMM which approximates the influence of groups of atoms or molecular fragments by treating them as single atoms.

van der Waals complexes

Molecular systems in which the individual parts are held together by non-covalent van der Waals forces.

Walsh diagram; Walsh-Mulliken diagram

A molecular orbital diagram where the orbitals in one reference geometry are correlated in energy with the orbitals of the deformed structure.

Walsh rules

The summaries of observations that the shapes of molecules in a given structural class are determined by the number of valence electrons.

Wax

A biological molecule that is an ester of a long-chain alcohol or a long-chain carboxylic acid.

Wigner rule

Wigner rule (also known as spin-conservation rule)—during an elementary chemical step, electronic and nuclear magnetic moments conserve their orientation.

Appendix 9: Terms Used in Analytical and Physical Chemistry

Ab initio; ab-initio; non-empirical

A quantum chemical calculation or prediction that is based purely on quantum theory rather than on experimental data. Accurate ab initio predictions are an important tool in modem chemistry, ab initio (Latin) = from first principles.

Absolute electronegativity

An absolute scale of electronegativity.

Absolute error; absolute uncertainty

The uncertainty in a measurement, expressed with appropriate units.

Absorption spectrum

Absorption of electromagnetic radiation by atoms (or other species) resulting from transitions from lower to higher energy states.

Activation analysis; neutron activation analysis

A technique in which a neutron, charged particle, or gamma photon is captured by a stable nuclide to produce a different, radioactive nuclide which is then measured.

Acceptor number (AN); GUTMANN

A quantitative measure, devised by GUTMANN (1976), of Lewis acidity.

Accuracy

The closeness of a single measurement to its true value, or the correctness of a single measurement.

Acid-base titration curve

A plot of the pH of a solution of acid (or base) against the volume of the added base (or acid).

Actino-

A prefix meaning "ray", "beam" used to form words such as actinochemistry (same as photochemistry) and actinometer, an instrument which measures the intensity of radiant energy.

Actinometer

Any instrument used to measure the intensity of radiant energy, particularly that of the sun.

Actinometry

The science of measurement of radiant energy, particularly that of the sun, in its thermal, chemical, and luminous aspects.

Activated complex; transition state

A highly reactive assembly of atoms, formed from reactant species.

Activation energy

The minimum energy of collision required for two molecules to react.

Active site; active center

Those sites for adsorption which are the sites for a particular heterogeneous catalytic reaction.

Activity (a)

An effective concentration or pressure to deviating from ideal behavior.

Adhesion

The process of fusing (binding) together of two surfaces that are normally separate.

Adiabatic (Thermodynamics)

A process occurring in a thermally insulated system.

Adiabatic calorimeter

Measuring enthalpy changes in chemical processes under adiabatic conditions where no flow of heat is allowed.

Adiabatic process

A process in which no exchange of heat between system and surroundings takes place.

Adiabatic system

One for which no exchange of heat between the system and surroundings is possible.

Adsorbate

The material being adsorbed.

Adsorbent

The material doing the adsorbing.

Adsorption

The process by which layers of a gas, liquid or solid build up on a surface.

Adsorption isotherm

A relationship between the partial pressure (or concentration) of an adsorbate substance and the surface coverage of the adsorbent at constant temperature.

Aerosol

A system of colloidal particles dispersed in a gas.

Affinity chromatography

A separation technique.

Aliquot

A measured portion of a sample taken for analysis.

American Society for Testing and Materials (ASTM)

A not-for-profit organization that develops and provides voluntary consensus standards, related I technical information, and services having internationally recognized quality and applicability.

Ampere (A); amp

The SI base unit of electric current.

Amperometry; amperometric

Determining the concentration of a material in a sample by measuring electric current.

AMSOL

A semiempirical quantum chemistry program.

Anisotropic

Non-isotropic.

Anode

The electrode at which oxidation occurs in a cell.

Assisted model building with energy refinement (AMBER)

A molecular mechanics modeling method.

Atmospheric chemistry

The study of the chemical constituents of the Earth atmosphere.

Atomic emission spectrum

A spectrum of radiation emitted by excited atoms.

Aufbau principle

A principle stating that as protons are added one by one to the nucleus to build up the elements, electrons are similarly added to hydrogen-like orbitals.

Auger electron spectroscopy (AES)

A chemical surface analysis method.

Auto-ignition temperature

The lowest temperature at which a material will ignite without an external source of ignition.

Azeotrope; azeotropic mixture; azeotropy

A solution that does not change composition when distilled.

Azeotropic drying

A method of removing water from a liquid.

Back donation

A description of the bonding of pi-conjugated ligands to a transition metal.

Balmer series

An equation which describes the emission spectrum of hydrogen.

Band theory

A theory that accounts for the bonding and properties of metallic solids.

Barometric distribution law

A law governing the distribution of gas molecules in the atmosphere as a function of their molar mass, height, temperature and the acceleration due to gravity.

Basis function

A mathematical function.

Basis set

A set of mathematical functions.

Bathochromic shift

Shift of a spectral band to lower frequencies (longer wavelengths) owing to the influence of substitution or a change in environment.

Battery

A group of galvanic cells connected in series.

Beer's law; Lambert-Beer law

An equation expressing the effect of concentration of the absorption medium on the absorption of electromagnetic radiation.

Bethe lattices

The infinite connected graphs not containing cycles, all vertices of which are equivalent and have equal numbers («) of neighbors.

Binding energy (molecular)

The difference between the total energy of a molecular system and the sum of the energies of its isolated p- and s-bonds.

Birefringence (double refraction)

An optical property some materials possess where double refraction occurs.

Bloch orbital

See Crystal orbital.

Bohr magneton; electronic Bohr magneton (μB)

A physical constant of magnetic moment of electrons; the magnitude of the magnetic dipole moment of an orbiting electron in the ground state.

Bolometer

An instrument which measures the intensity of radiant energy by employing a thermally sensitive electrical resistor; a type of actinometer.

Boltzmann equation

A statistical definition of entropy.

Born-Oppenheimer (BO) approximation

Separation of the total wavefunction of a molecular system into electronic and nuclear parts.

Boson (Quantum theory)

An elementary particle with integral or zero spin angular momentum quantum number.

Boyle's law

The pressure of a fixed amount of gas, at constant temperature, varies inversely with the volume.

Brackett series (Brackett)

The series which describes the emission spectrum of hydrogen when the electron is jumping to the fourth orbital. All of the lines are in the infrared portion of the spectrum.

Bulk modulus

Measuring the substance's resistance to compression.

Chemometrics

Chemometrics is the application of statistics to the analysis of chemical data (from organic, analytical or medicinal chemistry) and design of chemical experiments and simulations.

Canonical molecular orbitals; self-consistent field orbitals

The molecular orbitals which produce a matrix in the canonical (diagonal) form.

Capillary

A very narrow channel or tube having a small inner diameter through which a fluid can pass.

Capillary action; capillarity

The ability of a narrow tube to draw a liquid upwards against the force of gravity.

Capillary drying

The evaporation of moisture from the surface of a porous mass.

Carnot cycle

A hypothetical heat engine.

Carnot's theorem

No engine operating between two temperatures can be more efficient than a reversible engine.

Characteristic polynomial

A secular polynomial.

CHARMM (Molecular mechanics)

A program for macromolecular simulations, including energy minimization, molecular dynamics and Monte Carlo simulations.

Chemical potential

The molar Gibbs energy of a substance.

Chemical shift (NMR)

A parameter measuring the variation of the resonance frequency of a nucleus in NMR in consequence of its magnetic environment.

Chemiluminescence

A chemical reaction that releases energy as electromagnetic radiation.

Chemisorption

Adsorption in which the adsorbate is chemically bonded to the adsorbent.

Chemometrics

The application of statistics to the analysis of chemical data (from organic, analytical or medicinal chemistry) and design of chemical experiments and simulations.

Chromatography

A method for separating mixtures based on differences in the speed at which they migrate over or through a stationary phase.

Circular dichroism

An optical property of materials.

Clausius-Clapeyron equation

An equation that expresses the relation between the vapor pressure of a liquid and the absolute temperature.

CLogP values

Are calculated 1-octanol/water partition coefficients.

Closed shell systems

Even-electron atomic or molecular systems whose electron configurations consist of doubly occupied orbitals.

Coagulation

The process by which the dispersed phase of a colloid is made to aggregate and thereby separate from the continuous phase.

Combustion; combustion reaction

A chemical reaction between a fuel and an oxidizing agent that produces heat (and usually, light).

Common ion effect

The shift in an ionic equilibrium caused by the addition of a solute that provides an ion that takes part in the equilibrium.

Compton effect

Demonstrates that photons have momentum.

Computational chemistry

A branch of chemistry concerned with the prediction or simulation of chemical properties, structures, or processes using numerical computational techniques.

Computer-assisted drug design

Computer-assisted techniques used to discover, design and optimize biologically active compounds.

Computer-assisted molecular design (CAMD)

Involves all computer-assisted techniques used to discover, design and optimize compounds with desired structure and properties.

Computer-assisted molecular modeling (CAMM)

The investigation of molecular structures and properties using computational chemistry and graphical visualization techniques.

Concerted process

Two or more primitive changes are said to be concerted (or to constitute a concerted process) if they occur within the same elementary reaction.

Conformational analysis

The exploration of energetically favorable spatial arrangements (shapes) of a molecule (conformations) using molecular mechanics, molecular dynamics, quantum chemical calculations or analysis of experimentally-determined structural data.

Connolly surface

The envelope traced out by the point of contact of a defined probe (e.g., a sphere) and a molecule of interest.

Conservation of orbital symmetry

The orbital symmetry control of concerted reactions.

Correlation energy (Quantum chemistry)

The difference between the energy of a system calculated as the minimal value within the

Hartree-Fock approximation and the exact nonrelativitistic energy of that system.

Corrosion

The process by which metals are oxidized in the atmosphere.

Coulomb (C)

The SI unit of electric charge.

Coulomb repulsion

The potential energy component corresponding to the electrostatic interaction between two similarly charged particles.

Coupled cluster (CC) method

An *ab initio* quantum mechanical method in which electron correlation effects are incorporated by an exponential operator which acts on the zero-order wavefunction.

Craig plot

A plot of two substituent parameters.

Crystal field

The average static electric field experienced by an ion, molecule or atom in a crystal generated by all the other surrounding atoms, molecules, or ions.

Crystal field theory

A theory used to describe the electronic structure of transition metal complexes.

Crystal orbital; band orbital

A one-electron function extended throughout a crystal.

Crystalline solid; crystal

A solid in which the constituent atoms, molecules, or ions are packed in a regularly ordered, repeating pattern extending in all three spatial dimensions.

Crystallography

The determination and characterization of the structure of crystalline materials, typically through the use of analytical instruments.

Dalton's law of partial pressures

For a mixture of gases in a container, the total pressure exerted (observed) is equal to the sum of the pressures that each individual component gas would exert had it alone occupied the container at the same temperature.

Dative bond

The coordination bond formed upon interaction between molecular species one of which serves as a donor and the other as an acceptor of the electron pair to be shared in the complex formed.

Degenerate rearrangement

See Automerization.

Delocalization

A quantum mechanical concept to describe the pi bonding in a conjugated system.

Diamagnetic substances

Substances having a negative magnetic susceptibility and are repelled out of a magnetic field.

Dichroism

An optical property of crystals and solutions.

Differential scanning calorimetry (DSC)

A thermogravimetric technique by which the temperature of a sample of the substance in question is raised in increments while a reference is heated in the same rate.

Differential thermal analysis (DTA)

A thermogravimetric technique that is often used to analyze materials that react or decompose at higher temperatures.

Dimroth-Reichardt ET parameter

A measure of the ionizing power of a solvent.

Dispersion force; London force

An intermolecular attractive force that arises from a cooperative oscillation of electron clouds on a collection of molecules at close range.

Dobson units (DU) (ozone layer)

A measurement of column ozone levels.

Dosimeter

An instrument for measuring the ultraviolet in solar and sky radiation; also a device, worn by persons working around radioactive material, which indicates the dose of radiation to which they have been exposed.

Dry cell battery

A common battery used in calculators, watches, radios, and tape players.

Dynamic viscosity

The ratio of the shear stress to the velocity.

Effective nuclear charge

The positive charge that an electron experiences from the nucleus; equal to nuclear charge but reduced by shielding or screening from any other intervening electrons.

Electrical conductivity; electrical conductance; conductance

The ability to conduct an electric current; a measure of how easily an electric current can pass through a material.

Electrochemical cell

A device that uses a redox chemical reaction to produce electricity, or a device that uses electricity to drive a redox chemical reaction in the desired direction.

Electrochemistry

A branch of physical chemistry dealing with the study of the interchange of chemical and electrical energy.

Electrode

An electrically conducting surface that allows electrons to be transferred between reactants in an electrochemical cell.

Electrode potential

A potential difference between an electrode and its solution.

Electrolysis

A chemical change resulting from electric current.

Electrolyte

A substance that dissociates fully or partially into ions when dissolved in a solvent, producing a solution that conducts electricity.

Electromagnetic

Of or pertains to magnetism produced by or associated with electricity.

Electromagnetic field

A field of force associated with a moving electric charge and consisting of electric and magnetic fields that are generated at right angles to each other.

Electromagnetism

A branch of physics which studies electric and magnetic fields and electromagnetic radiation in particular.

Electron microscope

A microscope that uses a source of electrons to attain high magnification levels.

Electron spectroscopy for surface analysis (ESCA)

An analytical technique.

Electronic chemical potential

The quantity that measures the escaping tendency of electrons from a species in its ground state.

Electronic configuration

The allocation of electrons within an atom or a molecule to a set of correspondingly atomic or molecular orbitals complying with the Pauli exclusion principle.

Electronic state

An arrangement allowed by the laws of quantum mechanics of electrons within an atom, molecule (or system of molecules).

Electrophoresis

A technique used for separation of ions by differences in rate and direction of migration under the influence of an applied electric field.

Emission spectrum

A spectrum associated with emission of electromagnetic radiation by atoms (or other species) resulting from transitions from higher to lower energy states.

Emission spectrum of the hydrogen atom

A spectrum of radiation emitted by excited hydrogen atoms.

Emulsifier

A substance which stabilizes an emulsion.

Emulsion

A colloidal suspension of two immiscible (unblendable) liquids.

Endergonic

Having a positive standard Gibbs energy.

Endoergic

A reaction which consumes energy.

Endothermic

Refers to a reaction in which energy (as heat) flows into the system.

Energy dispersive spectroscopy (EDS)

A spectroscopic technique.

Energy gradient; gradient norm

First derivatives of the total energy with respect to nuclear coordinates.

Energy hypersurface

Synonymous with potential energy surface.

Enthalpy (H)

A thermodynamic state function.

Enthalpy of solution; heat of solution

The amount of heat that is absorbed or released when a substance enters solution.

Entropy (S)

A thermodynamic state function.

Enzyme

A large molecule (macromolecule), usually a protein, that catalyzes biological reactions by increasing the reaction rate.

Equalization of electronegativity

The principle states that in a molecule all the constituent atoms should have same electronegativity value, which would be the geometric mean of the electronegativities of isolated atoms.

Equilibrium geometry

Molecular geometry that corresponds to the true minimum on the respective potential energy surface.

ESR; EPR

Electron spin resonance; electron paramagnetic resonance.

Exchange repulsion; exchange integral

A quantum mechanical integral involving atomic or molecular orbitals.

Excimer

A dimer stable only in the electronically excited state formed by the interaction of an excited molecular entity with a ground state partner of the same structure.

Excited state

An electronic state other than the lowest energy state of a system.

Exergonic

Having a negative standard Gibbs energy.

Exoergic

A reaction which releases energy.

Exothermic

Refers to a reaction in which energy (as heat) flows out of the system.

Faraday

Laws of electrolysis.

Faraday constant

The electric charge carried by one mole of electrons.

Fermi hole

A region around an electron where the probability of finding another electron with the same spin is very small due to the action of the antisymmetry principle.

Fermi level

The average for the highest occupied and the lowest unoccupied levels in metals, semiconductors, etc.

Ferromagnetic

A material having the property of ferromagnetism.

Ferromagnetism

Strong magnetization induced by weak field: a property of some substances.

First law of thermodynamics

The energy of the universe remains constant.

Flocculation

The process in which small particles aggregate together to give bigger particles.

Fourier transform infrared spectroscopy (FTIR)

A vibrational spectroscopic analytical technique through infrared absorption by a substance.

Franck-Condon principle

The approximation that an electronic transition is most likely to occur without changes in the positions of the nuclei in a molecular entity and its environment.

Free valence

A reactivity index applied mostly to radical reactions of conjugated systems.

Freundlich isotherm

An adsorption isotherm.

Fuel cell

Galvanic cell for which the reactants are continuously supplied.

Fugacity (f)

An effective pressure, adjusted to take into account the non-ideality of a real gas.

Galvanic cell

A device in which chemical energy from a spontaneous oxidation-reduction reaction is changed to electrical energy that can be used to do work.

Galvanizing

A process in which steel is coated with zinc to prevent corrosion.

Gamma rays; gamma radiation

A very high energy form of electromagnetic radiation.

Gaussian type orbital (GTO)

An exponential function centered on an atom.

Geiger counter

A radiation detector.

Gel

A sol in which the solid particles fuse or entangle to produce a rigid or semirigid mixture.

Gel permeation chromatography

A separation technique.

Gel point

The stage at which a liquid begins to take on the semisolid characteristics of a gel.

Generalized valence bond (GVB) method

A multiconfigurational method that uses a limited set of valence bond configurations in the conventional multiconfiguration SCF method.

Gibbs energy (G); Gibbs free energy

A thermodynamic property or state function.

Gibbs-Helmholtz equation

An equation describing the effect of temperature on the change in Gibbs energy in a chemical process.

Glass-ceramic

A crystalline product created by the controlled crystallization of glass.

Graham's law of effusion

The rate of effusion of a gas is inversely proportional to the square root of its density.

Greenhouse effect

A warming effect exerted by certain molecules in the earth's atmosphere.

Greenhouse gases

Gases that absorb and trap heat in the atmosphere.

Ground state

The lowest energy state for an atom or molecule.

Hamiltonian

The quantum mechanical operator for the total energy.

Hansch analysis

The quantitative relationship between the biological activity of a series of compounds and their physicochemical parameters.

Hartree-Fock limit

The lowest energy that would be obtained via the Hartree-Fock SCF procedure if there were no restrictions on the sorts of function that molecular orbitals could adopt.

Hartree-Fock self-consistent field (HF-SCF)

Method for determination of the spatial orbitals of the many-electron determinantal wavefunction based on reducing coupled nonlinear differential equations by use of the variational method.

Heat of formation; standard enthalpy of formation

The amount of enthalpy involved in the formation of 1 mol of substance in its standard state from the elements in their standard state.

Heisenberg uncertainty principle

A principle stating that there is a fundamental limitation to how precisely we can know both the position and the momentum of a particle at a given time.

Helmholtz energy (A)

A thermodynamic state function.

Hess's law of heat summation

For a chemical equation that can be written as a sum of two or more steps, the enthalpy change for the overall equation equals the sum of the enthalpy changes for the individual steps.

Hessian matrix; force constant matrix

The matrix of second derivatives of energy.

Heterogeneous equilibrium

A state of chemical equilibrium that involves substances in different thermodynamic states.

Heterogeneous mixture

A mixture that has different properties in different regions of the mixture.

Heterogeneous reaction

A reaction involving reactants in different phases.

Heterolytic fission; heterolytic bond cleavage

The process in which a covalent bond is broken with one atom taking both electrons forming the bond.

High performance liquid chromatography (HPLC)

A separation technique.

High-spin state

A quantum state with the largest number of unpaired electrons.

HOMO

An acronym for highest occupied molecular orbital.

Homodesmotic reactions

A subclass of isodesmic reactions in which reactants and products contain equal numbers of carbon atoms in corresponding states of hybridization.

Homogeneous equilibrium

An equilibrium system in which all reactants and products are in the same state.

Homogeneous mixture

A mixture that is the same throughout; a solution.

Homogeneous reaction

A reaction involving reactants in only one phase.

Humidity

The moisture content of air.

Hund's rule

A rule for building up the electronic configuration of atoms and molecules.

Hückel molecular orbital (HMO) theory

The simplest molecular orbital theory of pi-conjugated molecular systems.

Hydrogen-like atomic orbitals

One-electron wavefunctions which are solutions of the Schroedinger equation for the one-electron atoms or ions, called hydrogen-like atoms.

Hydrophilic colloid

A hydrophilic colloid is a colloid in which there is a strong attraction between the dispersed phase and the continous phase (water).

Imbalance (chemical reaction)

The situation in which reaction parameters that characterize different bond forming or bond breaking processes in the same reaction have developed to different extents as the transition state is approached.

Infrared lamp

A special type of incandescent lamp that is designed to produce energy in the infrared portion of the electromagnetic spectrum.

Infrared radiation (IR)

An electromagnetic radiation.

Infrared spectroscopy (IRS)

A spectroscopic analytical technique using infrared absorption of a substance.

Inhibitor

A substance which is capable of stopping or retarding a chemical reaction.

Intensive property

A physical property whose magnitude does not depend upon the amount of material in a sample.

Intermolecular forces

Relatively weak interactions that occur between molecules.

Internal energy (U)

The sum of the kinetic and potential energies of all components of an object.

Intramolecular forces

Interactions that occur within a given molecule.

Ionization potential (atom or molecule)

The minimal energy needed for the removal of an electron from an atom or a molecule.

Isodesmic reaction

A balanced chemical reaction (actual or hypothetical) in which the types of bonds that are made in forming the products are the same as those which are broken in the reactants.

Isogyric reaction

A reaction in which the number of electron pairs in reactants and products is conserved.

Isotherm

At constant temperature.

Isotope effect (rate and equilibrium)

The effect on the rate or equilibrium constant of two reactions that differ only in the isotopic composition of one or more of their otherwise chemically identical components.

Isotope effect (heavy atom)

An isotope effect due to isotopes other than those of hydrogen.

Isotope exchange

A chemical reaction in which the reactant and product chemical species are chemically identical but have different isotopic composition.

Isotopic perturbation method

NMR shift difference measurement of the isotope effect on a fast (degenerate) equilibrium between two, except for isotopic substitution, species which are equivalent.

Isotopic scrambling

The achievement, or the process of achieving, an equilibrium distribution of isotopes within a specified set of atoms in a chemical species or group of chemical species.

Isotropic

Having identical properties along every direction.

Infrared lamp

A special type of incandescent lamp that is designed to produce energy in the infrared portion of the electromagnetic spectrum.

Isotopic perturbation method

NMR shift difference measurement of the isotope effect on a fast (degenerate) equilibrium between two, except for isotopic substitution, species which are equivalent.

Jahn-Teller effect

Deals with molecular distortions due to an electronically degenerate ground state.

Kinematic viscosity

A coefficient defined as the ratio of the dynamic viscosity of a fluid to its density.

Kinetic energy

The energy a body has by virtue of its motion.

kinetic isotope effect

The effect of isotopic substitution on a rate constant.

Kirchoff relationships

Equations relating the temperature dependence of enthalpies of reaction.

Klystron

An electron tube for converting direct-current energy into radio frequency energy by alternately speeding up and slowing down the electrons.

Koopmans' theorem

Relates experimental ionization potentials with energy levels of molecular orbitals.

Lambert's law; Bouguer's law; Lambert-Bouguer law

Relates the absorption of light to the properties of the material through which the light is traveling, and expresses the effect of thickness of the material.

Laminar flow

In fluid flow, a smooth flow in which no crossflow of fluid particles occur between adjacent streamlines; hence, a flow conceived as made up of layers—commonly distinguished from turbulent flow.

Langmuir adsorption isotherm

An adsorption isotherm.

Langmuir-Hinshelwood mechanism

A mechanism for heterogeneous catalysis.

Laser

An acronym for light amplification by stimulated emission of radiation.

Latent heat

Heat that is absorbed without causing a rise in temperature.

Law

A statement, usually mathematical, which describes some physical phenomena.

Ligand field

See Crystal field.

Line spectrum; line spectra; line emission spectrum

A spectrum showing only certain discrete wavelengths.

Linear combination of atomic orbitals (LCAO)

The approximation of the molecular orbital function as a linear combination of atomic orbitals chosen as the basis functions.

Linear free-energy relation; linear Gibbs energy relation

A linear correlation between the logarithm of a rate constant or equilibrium constant for one series of reactions and the logarithm of the rate constant or equilibrium constant for a related series of reactions.

Linear solvation energy relationships

Equations involving the application of solvent parameters in linear or multiple (linear) regression expressing the solvent effect on the rate or equilibrium constant of a reaction.

Liquid
A state of matter that has a high density and is incompressible compared to a gas.

Liquid crystals
Substances that exhibit a phase of matter that has properties between those of a conventional liquid, and those of a solid crystal.

Localized molecular orbitals (LMO)
The molecular orbitals located on certain fragments of a molecular system.

Low-spin state
A quantum state with the smallest number of unpaired electrons.

Lubricity (lubrication oil)
A measure of the ability of an oil or other compound to lubricate (reduce friction) between two surfaces in contact.

Luminescence
The process in which an atom or molecule emits a photon of light.

LUMO
An acronym for lowest unoccupied molecular orbital.

Magnetic field
A region of space wherein any magnetic dipole would experience a magnetic force or torque.

Magnetic field intensity; magnetic intensity
The magnetic force exerted on an imaginary unit magnetic pole placed at any specified point of space.

Magnetic induction
A measure of the strength of a magnetic field existing within a magnetic medium.

Magnetic moment
The quantity obtained by multiplying the distance between two magnetic poles by the average strength of the poles.

Magnetoelectric
Of or pertaining to electricity produced by or associated with magnetism.

Mass spectrometry; mass spectrometer
A technique for separating ions by their mass-to-charge (w/z) ratios.

Microscope
An instrument for viewing very small objects.

Microscopic
A physical entity or process of small scale.

Microscopy
A technique for producing visible images of structures or details too small to otherwise be seen by the human eye.

Microwave; microwave radiation
Electromagnetic radiation with wavelength between 3 mm and 30 cm.

Millikan oil drop experiment

Experiment designed to measure the electronic charge.

Miscible; miscibility; liquid miscibility

Two liquids are considered miscible or mixable if they are mutually soluble in all proportions.

Miscible liquids

Liquids that mix with or dissolve in each other in all proportions.

Mixture

A material of variable composition that contains two or more substances.

Moderator

A substance used in a nuclear reactor to slow down the neutrons.

Molar enthalpy of formation; standard enthalpy of formation

The amount of enthalpy involved in the formation of 1 mol of substance in its standard state from the elements in their standard state.

Molar heat capacity (C_m)

The amount of heat required to raise the temperature of one mole of a material by 1 degree celsius.

Molar heat of fusion

The energy required to melt 1 mol of a solid.

Molar heat of vaporization

The energy required to vaporize 1 mol of a liquid.

Molecular descriptors

Terms that characterize a specific aspect of a molecule.

Molecular dynamics (MD)

The computation of the motion of atoms within a molecular system using molecular mechanics.

Molecular entity

Any constitutionally or isotopically distinct atom, molecule, ion, ion pair, radical, radical ion, complex, conformer *etc.*, identifiable as a separately distinguishable entity.

Molecular graphics

The visualization and manipulation of three-ft dimensional representations of molecules on a graphical display device.

Molecular modeling

A technique for the investigation of molecular structures and properties using computational chemistry.

Molecular orbital

A wavefunction which depends explicitly on the spatial coordinates of only one electron and extends over two or more atoms in a molecule.

Molecularity (Chemical kinetics)

The number of reactant particles (molecules, atoms, or ions) that are involved in an elementary reaction.

Monochromatic

Radiation that has a single wavelength or more realistically a radiation with a very narrow range of wavelengths.

Monodisperse colloidal system

A colloidal system having particles of the same size, interaction, and shape.

Monte Carlo simulation

Simulation methods that use random numbers to generate possible molecules or materials and then identify the optimal system, for example through molecular mechanics.

Morse potential; Morse function

The empirical function relating the potential energy of a molecule to the interatomic distance.

Mössbauer effect

The phenomenon in which an atom in a crystal undergoes no recoil when emitting a gamma ray, giving all the emitting energy to the gamma ray, resulting in a sharply defined spectral lines (narrow wavelength).

Mulliken absolute electronegativies

An absolute scale of electronegativity.

Moller-Plesset perturbation theory; Moller- Plesset (MP)

An approach to account for electron correlation.

Nephelometric

Method of measuring turbidity.

Network solid

An atomic solid containing strong directional covalent bonds.

Neutralization reaction

An acid-base reaction.

Nickel-cadmium cell

A voltaic cell.

Nuclear binding energy

Energy needed to break an atomic nucleus into separate protons and neutrons.

Nuclear magnetic resonance (NMR)

An analytical technique based on magnetic resonance of nuclei.

Number of significant figures

The number of digits reported for the value of a measured or calculated quantity, indicating the precision of the value.

Ohm's law

$V = IR$, where V is the potential across a circuit element, I is the current through it, and R is its resistance.

Open-shell systems

Atomic or molecular systems in which the electrons are not completely assigned to orbitals in pairs.

Optical isomers; enantiomers

Isomers that are nonsuperimposable mirror images of one another.

Optically active

A substance having the ability to rotate the plane of polarized light.

Orbital interaction

Interaction of two orbitals due to their overlap which results in the formation of two new orbitals.

Osmotic pressure

Pressure which must be applied to a solution to prevent water from flowing in via a semipermeable membrane.

Overall order of reset Ion (Chemical kinetics)

The sum of the exponents in the rate law.

Oxidation half reaction

That part of a redox reaction that involves loss of electrons.

Pairing energy (Crystal field theory)

The energy required to put two electrons into the same orbital.

Paramagnetic

A substance that is weakly magnetized so that it will lie parallel to a magnetic field.

Pariser-Parr-Pople (PPP) method

A semiempirical quantum mechanical method of calculation of the properties of conjugated molecules and ions from self-consistent-field theory and the p-electron approximation.

Parity

A symmetry property of a wave function.

Partial pressures

The independent pressures exerted by different gases in a mixture.

Particle accelerator

A device used to accelerate nuclear particles to very high speeds.

Paschen series

The series which describes the emission spectrum of hydrogen when the electron is jumping to the third orbital. All of the lines are in the infrared portion of the spectrum.

Pattern recognition

The identification of patterns in large data sets using appropriate mathematical methodologies.

Pauli exclusion principle

No two identical fermions in a system, such as electrons in an atom, can have an identical set of quantum numbers.

Permutational isomerization

See Automerization.

Perturbation theory

A quantum-mechanical approximation method.

Pfund series

The series which describes the emission spectrum of hydrogen when the electron is jumping to fifth orbital. All of the lines are in the infrared portion of the spectrum.

Pharmacokinetics (Pharmacology)

A branch of pharmacology.

Photoconductor

A material that becomes a good conductor when illuminated with light.

Photoelectric effect

The emission of an electron from a surface as the surface absorbs a photon of electromagnetic radiation; electrons so emitted are termed.

Photoelectrons

An electron which has been ejected from its parent atom by interaction between that atom and a high-energy photon.

Photoelectron spectroscopy (PES)

A spectroscopic technique involving the bombardment of a sample with radiation from a high-energy monochromatic source and the subsequent determination of the kinetic energies of the ejected electrons.

Photoionization

The ionization of an atom or molecule by the collision of a high-energy photon with the particle.

Photometer

An instrument for measuring the intensity of light or the relative intensity of two beams of lights.

Photometry

The study of the measurement of the intensity of light.

Physical property

A characteristic of a substance.

Plank's constant (h)

A physical constant that is used to describe the sizes of quanta.

Poisson ratio (v)

Ratio of the strain in the direction of the applied load to the strain normal to the load.

Positron

A subatomic particle that is similar to an electron and has the same mass but a positive charge.

Potential energy

Energy due to position or composition.

Potential energy curve

A curve that shows the potential energy of a reaction against the progress of the reaction.

Potential energy surface (PES)

In quantum mechanics, the total energy of an arrangement of atoms can be represented as a curve or (multidimensional) surface, with atomic positions as variables.

Precision

The closeness of the set of values obtained from identical measurements of a quantity.

Predissociation

Dissociation of a molecule occurring through its excitation to a certain bound state which is coupled to a dissociative continuum.

Qualitative analysis

The determination of the identity of substances present in a sample.

Quantitative analysis

The determination of the amount of a substance or species present in a sample of material.

Quantitative structure-activity relationship (QSAR)

A mathematical model that relates a quantitative measure of chemical structure to a physical property or to a biological effect.

Quantum; quanta

A discrete packet of energy.

Quantum mechanics

A branch of physics that describes the behavior of objects of atomic and subatomic size.

Quantum number

A numerical label used to classify a quantum state.

Quantum theory

A fundamental physical theory.

Rad

The dosage of radiation that deposits 1×10^{-2} Joule of energy per kilogram of tissue.

Radian (rad)

An angle with vertex at the center of a circle of radius r that encompasses an arc of length r.

Radioactive decay (radioactivity)

The spontaneous disintegration of a nucleus.

Radioactivity

Spontaneous emission of radiation from unstable (radioactive) elements.

Radiocarbon dating (carbon-14 dating)

A method for dating ancient objects on the basis of the radioactive decay.

Radiotracer; radioactive tracer

A very small amount of radioactive nuclide (isotope) introduced into a chemical, physical or biological system.

Random error

An error that has an equal probability of being high or low.

Raoult's law

The vapor pressure of the solution is directly proportional to the mole fraction of the solvent in the solution.

Rate determining step

The slowest step in a reaction mechanism.

Rate law

An equation that relates the rate of a reaction to the concentrations of reactants raised to various powers.

Rate of decay

The change per unit time in the number of radioactive nuclides in a sample.

Reaction dynamics

A branch of chemical kinetics.

Reaction order

The exponent of the concentration of a given reactant species in the rate law, as determined experimentally.

Reaction quotient (Q)

An expression that has the same form as the equilibrium constant expression but whose concentration values are not necessarily those at equilibrium.

Reaction rate

The change in molar concentration of a product or reactant per unit time.

Refraction

The process in which the direction of energy propagation is changed as the result of a change in density within the propagating medium.

Refractive index

The ratio of the velocity of radiation in a vacuum to its velocity in the medium through which it passes.

Relativistic effects

Corrections to exact nonrelativistic energy from the fact that inner shell electrons in heavy atoms move with velocities comparable in order of magnitude to the velocity of light.

Rem (units)

A unit of radiation dosage.

Reversible processes

Processes which may be made to proceed in the forward or reverse direction by the (infinitesimal) change.

Rigidity modulus; sheer modulus

The ratio of shear stress to the shear strain.

Saddle point

A point of lowest maximum energy on a valley (reaction path) connecting two minima on the potential energy surface.

Salt bridge

A U-tube containing an electrolyte.

Scalar

A quantity that has no direction in space, only an amount, such as the mass of an object.

Scanning electron microscope (SEM)

A type of electron microscope with scanning capability.

Scanning electron microscopy (SEM)

Scan images at magnifications of 15-300,000×.

Scanning probe microscopy (SPM); atomic force microscopy (AFM)

A scanning probe microscopy technique.

Scanning tunneling microscope (STM)

A scanning probe microscopy instrument used to obtain images of conductive surfaces at an atomic scale.

Schrodinger (Schroedinger) wave equation

An equation used in quantum mechanics to describe the behavior of a system of microscopic particles.

Scintillation counter

An instrument that measures radioactive decay.

Second law of thermodynamics

It is impossible for an engine to perform work by cooling a portion of matter to a temperature below that of the coldest part of the surroundings.

Semipermeable membrane

A selectively permeable membrane.

Sigma, pi molecular orbitals

The terms that are symmetry designations of molecular orbitals.

Signal-to-Noise ratio

The ratio of one parameter of a wanted signal to the same parameter of the noise at any point in an optical spectrum, electrical circuit, device, or transmission system.

Significant figures

Digits that contribute to the accuracy of a measurement.

Slater determinant

The determinantal representation of many-electron wavefunction which conforms to the requirement of the antisymmetry principle.

Slater type atomic orbital (STO)

The exponential function centered on an atom.

Solar radiation

Radiant energy emitted by the sun.

Solvation energy

The change in Gibbs energy when an ion or molecule is transferred from a vacuum (or the gas phase) to a solvent.

Solvent isotope effect

A kinetic or equilibrium isotope effect resulting from change in the isotopic composition of the solvent.

Specific heat capacity

The amount of heat required to raise the temperature of a unit mass of a material by 1 degree celsius.

Spin crossover

A type of molecular magnetism that is the result of electronic instability caused by external constraints (temperature, pressure, or electromagnetic radiation).

Spin density

The excess of the electron density related to the electron with a spin over that of the electron with b spin (see spin-orbital) at a given point of an open-shell system. For a closed-shell system spin density is zero everywhere.

Spin polarization

Static and dynamic spin polarization effects are distinguished.

Spin projection

A component of the angular spin moment along an arbitrary axis (usually chosen as the z-direction).

Spin quantum number (ms)

The quantum number that specifies the direction in which the electron is spinning with values of $^+1/2$ and $-1/2$.

Spin-coupled (SC) wavefunction

Representation of a wavefunction in the modified valence bond theory.

Spin-orbit coupling

The interaction of the electron spin magnetic moment with the magnetic moment due to the orbital motion of the electron, and the consequent mixing of electronic states of different multiplicity.

Spin-orbital

The complete one-electron wavefunction given by a product of a spatial function and a spin function.

Spin-spin coupling

A small relativistic effect due to interaction between the spin magnetic moments of electrons or nuclei.

Spontaneous process

A physical or chemical change that occurs by itself (without external influence).

Standard deviation (Statistics)

A measure of statistical dispersion.

Standard molar entropy ($S^\ominus$)

The entropy of one mole of a substance in its standard state.

Standard reduction potential ($E^\ominus$)

The voltage associated with a reduction process at standard state.

Standard solution

A solution of precisely known concentration.

Standard uncertainty (in measurement)

Uncertainty of the result of a measurement expressed as a standard deviation.

Stationary point (quantum)

A point on the potential energy surface at which the energy gradients with respect to all coordinates vanish.

Stationary state (quantum)

A quantum state of a system in which the expectation values of properties do not change with time.

Stereochemical nonrigidity

The capability of a molecule to undergo fast and reversible intramolecular isomerization.

Steric effect; steric hindrance

The effect on a chemical or physical property (structure, rate or equilibrium constant) upon introduction of substituents having different steric requirements.

Steric isotope effect

A secondary isotope effect attributed to the different vibrational amplitudes of isotopologues.

Stopped flow (Kinetics)

A technique for following the kinetics of reactions in solution (usually in the millisecond time range).

Strain energy; steric energy

The excess energy due to steric strain of a molecular entity or transition state structure.

Structure-activity relationship (SAR)

A (qualitative) association between a chemical substructure and the potential of a chemical containing the substructure to exhibit a certain biological effect.

Structure-based design

A design strategy for new chemical entities based on the three-dimensional (3D) structure of the target.

Structure-property correlations

Refers to all statistical mathematical methods used to correlate any structural property to any other property.

Superacid

A medium having a high acidity.

Superconductivity

A phenomenon in which a material, when cooled below its superconducting transition temperature T_c.

Supercooling

The process of cooling a liquid to a temperature below its freezing point without its changing to a solid.

Superheating

The process of heating a liquid to a temperature above its boiling point without its boiling.

Supermolecule

A discrete oligomolecular species that results from the intermolecular association of its components.

Suramolecular chemistry

A field of chemistry related to species of greater complexity than molecules, which are held together and organized by means of intermolecular interactions.

Surface tension

The energy required to increase the surface area of a liquid by a unit amount.

Synchronization (principle of nonperfect synchronization)

Applies to reactions in which there is a lack of synchronization between bond formation or bond rupture and other primitive changes that affect the stability of products and reactants.

Systematic error

An error that always occurs in the same direction.

Standard uncertainty (in measurement)

Uncertainty of the result of a measurement expressed as a standard deviation.

Tesla (T)

The derived SI unit of magnetic flux density.

The solvent-accessible surface

The surface traced out by of a probe molecule, e.g., water, rolling over the van der Waals surface of a molecule.

Thermal radiation

The electromagnetic radiation emitted by any substance as the result of the thermal excitation of its molecules Thermal radiation ranges in wavelength from the longest infrared radiation to the shortest ultraviolet radiation.

Thermochemical equation (Thermochemistry)

A balanced chemical equation with full information about the physical states of chemicals, and the energy (enthalpy) changes involved in the reaction.

Thermochemistry

The study of enthalpy (energy) changes in chemical processes.

Thermodynamics; chemical thermodynamics

The study of energy changes that accompany chemical and physical processes.

Third law of thermodynamics

The entropy of any pure crystalline substance is zero at the absolute zero.

Thixatropic gel

A gel which reverts, under isothermal conditions, to the sol state under the influence of motion.

Thixotropy

The (isothermal) process in which the viscosity of a substance decreases as the substance is set in motion by some mechanical action such as stirring or shaking.

Through-bond interaction

An intramolecular orbital interaction of spatially separated orbitals, where the orbitals interact through their mutual mixing with sigma-orbitals of the intervening framework.

Through-space interaction

An orbital interaction that results from direct spatial overlap of two orbitals.

Topomerization

See Automerization.

Total energy of a molecule

The sum of the total electronic energy and the energy of internuclear repulsion.

Transition state theory (TST)

The theory which provides a conceptual framework for understanding all chemical reactivity.

Tyndall effect

The observance of the path of a beam of light through a liquid due to the scattering of light by dispersed colloidal-size particles.

Uncertainty (in measurement)

Characterizes the dispersion of the measured values.

Ultraviolet radiation (UV)

An electromagnetic radiation with wavelengths shorter than visible light.

Unit cell

The smallest unit from which of a crystal is constructed by stacking the units in three dimensions.

UVA

A band of ultraviolet radiation with wavelengths from 320~400 nanometers.

UVB

A band of ultraviolet radiation with wavelengths from 280~320 nanometers.

UVC

A band of ultraviolet radiation with wavelengths shorter than 280 nanometers.

van der Waal's equation of state

An empirical equation of state of a real gas.

Vaporization (evaporation)

The change in state that occurs when a liquid evaporates to form a gas.

Variational principle; variational method

The quantum mechanical energy calculated from Schroedinger equation using an approximate wavefunction will always yield an energy that is higher than the actual energy of the system.

Vector

A quantity that has both an amount (magnitude) and a direction in space.

Vibronic transition

A transition which involves a change in both the electronic and vibrational quantum numbers of a molecular entity. The transition occurs between two electronic states, but involves a change in both electronic and vibrational energy.

Virtual orbital

Any molecular orbital which is higher in energy than the HOMO level.

Viscosity

The resistance of a liquid to flow.

Vitreous

Refers to a material in a glassy amorphous state.

Volt(V)

The unit of measurement for electric potential.

Voltaic cell; galvanic cell

An electrochemical cell.

Volumetric analysis

A method of analysis based on titration such as acid-base and complexometric titration.

Wavefunction

A solution of Schrodinger equation.

Wigner rule

Also known as spin-conservation rule-during an elementary chemical step, electronic and nuclear magnetic moments conserve their orientation.

Work force

Acting over a distance.

X-ray photoelectron spectroscopy (XPS)

A chemical surface analysis method.

Young's modulus; modulus of elasticity

A measure of the stiffness of a material.

Zeeman effect; Zeeman line splitting

The splitting of the lines in a spectrum when the source is exposed to a magnetic field.

Zeolites

Natural hydrated alumino-silicates minerals that have a microporous structure.

References

[1] Atomic Structure[EB/OL]. (2016-06-15) [2019-04-21]. http://www.chemistryexplained.com/Ar-Bo/Atomic-Structure.html.

[2] Wikipedia, the Free Encyclopedia. Chemical engineering[EB/OL]. (2019-02-13) [2019-04-21]. https://en.wikipedia.org/w/index.php?title=Chemical_engineering&oldid=903396073.

[3] Wikipedia, the Free Encyclopedia. Chemistry[EB/OL]. (2019-01-08) [2019-04-21]. https://en.wikipedia.org/w/index.php?title=Chemistry&oldid=903416133.

[4] Wikipedia, the Free Encyclopedia. Atomic theory[EB/OL]. (2019-03-23)[2019-04-21]. https://en.wikipedia.org/w/index.php?title=Atomic_theory&oldid=907326979.

[5] Wikipedia, the Free Encyclopedia. Chemical bond[EB/OL]. (2018-12-13)[2019-04-21]. https://en.wikipedia.org/w/index.php?title=Chemical_bond&oldid=901325137.

[6] Wikipedia, the Free Encyclopedia. Chemical reaction[EB/OL]. (2019-02-22)[2019-04-21]. https://simple.wikipedia.org/w/index.php?title=Chemical_reaction&oldid=5992252.

[7] Encyclopædia Britannica, inc. Chemical reaction[EB/OL]. (2019-02-01)[2019-04-21]. https://www.britannica.com/science/chemical-reaction.

[8] WAGNER E C. Laboratory Methods of Organic Chemistry (Gattermann, L.; revised by Heinrich Weiland)[J]. Journal of Chemical Education, 1937, 14(7).

[9] F. A. BETTELHEIM, J M L. Laboratory Experiments for General, Organic, and Biochemistry[M]. Fort Worth: Harcourt College Publishers, 2000.

[10] F.W. FIFIELD, D K. Principles and Practice of Analytical Chemistry[M]. Oxford: Blackwell Science Ltd., 2000.

[11] HARDING A. Profile Sir Greg Winter—humaniser of antibodies[M]. Lancet, 2006.

[12] SlidesLive. Therapeutics Antibodies: A Revolution in Pharmaceuticals[EB/OL]. (2015-10-23) [2019-04-21].https://slideslive.com/38895318/therapeutics-antibodies-a-revolution-in-pharmaceuticals.

[13] ARNOLD F, MACUARE, K. A. The NAI Fellow Profile: An Interview with Dr. Frances Arnold[M]. Technology and Innovation, 2016.

[14] DEVINE P N, HOWARD R M, KUMAR R, et al. Extending the application of biocatalysis to meet the challenges of drug development [J]. Nature Reviews Chemistry, 2018, 2(12): 409-421.

[15] Chemistry World. In situ with Frances Arnold[EB/OL]. (2018-04-01)[2019-04-21]. https://www.

chemistryworld.com/culture/in-situ-with-frances-arnold/3008732.article.

[16] HARTINGS M R, AHMED Z. Chemistry from 3D printed objects[J]. Nature Reviews Chemistry, 2019, 3(5): 305-314.

[17] KEAN S. A storied Russian lab is trying to push the periodic table past its limits—and uncover exotic new elements[J]. Science, 2019.

[18] WALDMANN. Greg Winter: Pioneering Antibody Drugs[N], Medical Research Council, Mar. 18, 2013.

[19] 董坚. 化学化工专业英语[M]. 杭州：浙江大学出版社，2010.

[20] 刘宇红. 化学化工专业英语[M]. 北京：中国轻工业出版社，2000.

[21] 清华大学《英汉技术词典》编写组. 英汉技术词典[M]. 北京：国防工业出版社，1985.

[22] 邵荣，许伟，吕慧华. 新编化学化工专业英语[M]. 上海：华东理工大学出版社，2017.

[23] 张小军. 化工专业英语[M]. 北京：化学工业出版社，2015.